Recent Applications of Machine Learning in Natural Language Processing (NLP)

Recent Applications of Machine Learning in Natural Language Processing (NLP)

Guest Editors

Carlos A. Iglesias
Jinyi Zhang

MDPI

Basel • Beijing • Wuhan • Barcelona • Belgrade • Novi Sad • Cluj • Manchester

Guest Editors

Carlos A. Iglesias
Intelligent Systems Group
Universidad Politécnica de Madrid
Madrid
Spain

Jinyi Zhang
School of Information Science and Engineering
Shenyang Ligong University
Shenyang
China

Editorial Office
MDPI AG
Grosspeteranlage 5
4052 Basel, Switzerland

This is a reprint of the Special Issue, published open access by the journal *Applied Sciences* (ISSN 2076-3417), freely accessible at: https://www.mdpi.com/journal/applsci/special_issues/R45GDXG1X7.

For citation purposes, cite each article independently as indicated on the article page online and as indicated below:

Lastname, A.A.; Lastname, B.B. Article Title. *Journal Name* **Year**, *Volume Number*, Page Range.

ISBN 978-3-7258-4533-0 (Hbk)
ISBN 978-3-7258-4534-7 (PDF)
https://doi.org/10.3390/books978-3-7258-4534-7

Contents

About the Editors

Carlos A. Iglesias

Carlos A. Iglesias is a Full Professor at the School of Telecommunications Engineering at the Technical University of Madrid, where he received his M.S. and Ph.D degrees. He has been Deputy Director of R&D at Grupo Gesfor and Director of Innovation at Germinus XXI, as well as Professor at the universities of Valladolid and Pontificia de Comillas in Madrid. His research interests focus on intelligent systems applied to telecommunications, agent-oriented software engineering, big data, and knowledge engineering. He has published over 120 research articles at conferences and in books and national and international magazines. He is a member of the Program Committee of the leading Agent-Oriented Software Engineering conference. He actively participates in national and European research projects, the most recent being the FP7 Romulus, FP7 Omelette, FP7 EuroSentiment, H2020 SmartOpenData, H2020 MixedEmotions, and H2020 Trivalent projects in the areas of linked data and big data, applied to mashup technology, semantic discovery, and sentiment analysis.

Jinyi Zhang

Jinyi Zhang is an Associate Professor at the School of Information Science and Engineering, Shenyang Ligong University, China, and currently a visiting researcher at Gifu University, Japan. He received his B.S. degree in Information Management and Information Systems from Dalian University of Foreign Languages in 2011 and his M.S. and Ph.D. degrees in Electronics and Information Systems Engineering from Gifu University in 2017 and 2020, respectively. His research focuses on natural language processing, particularly large language models and their applications in machine translation, knowledge-driven language understanding, and multimodal interaction. He has published more than 20 academic papers, including over 10 in SCI-indexed journals, and serves as a reviewer or program committee member for top-tier conferences such as NeurIPS, AAAI, ICLR, ACL, ICML, EMNLP, and NAACL. He is also an active member of the Chinese Information Processing Society and the Association for Natural Language Processing in Japan. His academic journey has been supported by national fellowships from the China Scholarship Council (CSC) for both doctoral and postdoctoral research in Japan. He has led and participated in multiple research projects funded by national, provincial, and industrial sources, including the Educational Department of Liaoning Province and Huawei.

Preface

This Reprint brings together a collection of recent studies on the use of machine learning techniques in natural language processing (NLP). The eight selected papers cover a variety of topics, including few-shot named entity recognition, topic evolution analysis based on patent data, sentiment analysis involving multimodal inputs, domain-specific language modeling, and the application of large models to ontology construction and financial question answering. Together, these works reflect the growing need for more adaptable and context-aware NLP systems across different domains.

In recent years, NLP has made remarkable strides, particularly with the emergence of large language models. However, the field continues to face challenges such as limited training data in specialized domains, the difficulty of integrating multimodal signals, and the need for more transparent and controllable model behavior. The aim of this Special Issue was to showcase novel research directions that respond to these challenges through improved model design, learning strategies, and application-oriented approaches. The contributing papers demonstrate practical advancements that go beyond conventional NLP pipelines and offer fresh perspectives on current research questions.

We extend our sincere thanks to all authors for sharing their important work and to the reviewers for their time and valuable comments. We also appreciate the support of the editorial team at *Applied Sciences*. It is our hope that this Reprint will serve as a helpful reference for readers who are interested in current developments at the intersection of NLP and machine learning.

Carlos A. Iglesias and Jinyi Zhang
Guest Editors

applied sciences

MDPI

Editorial

Special Issue on Recent Applications of Machine Learning in Natural Language Processing (NLP)

Jinyi Zhang [1,2,*] and Carlos Á. Iglesias [3]

1 School of Information Science and Engineering, Shenyang Ligong University, Shenyang 110159, China
2 Faculty of Engineering, Gifu University, Gifu 501-1193, Japan
3 Intelligent Systems Group, Universidad Politécnica de Madrid, 28040 Madrid, Spain; carlosangel.iglesias@upm.es
* Correspondence: zhangjinyi@sylu.edu.cn

1. Introduction

Machine learning has profoundly advanced natural language processing (NLP), enabling more intelligent and human-like interaction across various applications, such as sentiment analysis, entity recognition, syntax parsing, and machine translation. The rise of large language models (LLMs) has further transformed tasks such as question answering and multilingual communication. With the integration of multimodal data—text, speech, and vision—NLP systems continue to expand in scope and sophistication.

This Special Issue, entitled "Recent Applications of Machine Learning in Natural Language Processing (NLP)", presents eight peer-reviewed papers showcasing recent research progress at the intersection of machine learning and NLP. These contributions span key areas such as named entity recognition, multimodal sentiment analysis, domain-specific language modeling, and knowledge-enhanced methods.

2. Overview of Contributions

This Special Issue comprises eight papers, each outlining novel approaches and findings in applying machine learning to NLP tasks. Below is an overview of each paper: Xiao et al. [1] proposed a novel few-shot named entity recognition method by integrating large language models with metric learning, improving performance in low-resource settings. Jiang et al. [2] developed a dynamic topic modeling framework for tracking 6G technology trends using patent texts and LLM-enhanced summarization. Tian et al. [3] introduced a collaborative learning approach for video temporal grounding, achieving high accuracy by leveraging vision–language model interactions. Doumanas et al. [4] conducted a comparative study of GPT-4 and Mistral in ontology engineering tasks, demonstrating the strengths and trade-offs of each model. Cao et al. [5] presented a prompt-based NER system for shearer maintenance text, extracting entities in a specialized industrial domain without fine-tuning. Yang et al. [6] proposed a multi-task pre-training framework to enhance few-shot multimodal sentiment analysis by aligning textual and visual features. Iaroshev et al. [7] evaluated retrieval-augmented generation models for financial report question answering, showing the importance of retrieval quality and structure. Gao et al. [8] introduced a graph convolutional network enhanced with sentiment support for aspect-level sentiment classification, achieving improved accuracy.

3. Conclusions

In summary, the contributions in this Special Issue collectively illustrate the breadth and depth of current research on machine learning applications in NLP. The eight papers cover a range of essential problems and methodologies: from advancing few-shot learning techniques and prompt-based models for specialized information extraction, to leveraging multi-modal and collaborative learning for integrating vision and language, and enhancing language model capabilities for knowledge-rich tasks like ontology management and financial document Q&A. A unifying theme is the continual push to overcome data scarcity and complexity by cleverly utilizing large pre-trained models, innovative training strategies, or incorporating domain-specific knowledge. These students not only advance their respective fields but also underscore the versatility of machine learning in tackling diverse NLP challenges across domains (from social media and multimedia content to industrial and financial text analytics). We believe that the insights and results presented in this Special Issue will inspire further research and development in the NLP community. Open challenges remain, such as improving model interpretability, handling low-resource languages, and ensuring the ethical use of AI in language tasks, which future studies can build upon using the foundations laid by these papers.

Funding: This research received no external funding.

Acknowledgments: As Guest Editors, we thank all the authors for contributing their outstanding research to this issue and the reviewers for their thorough evaluations and constructive feedback. We also thank the editorial team of Applied Sciences for their professional support in making this Special Issue a success. We hope that readers find this collection informative and stimulating, and that it serves as a valuable resource for researchers and practitioners working at the exciting intersection of machine learning and natural language processing.

Conflicts of Interest: The authors declare no conflicts of interest.

References

1. Xiao, Y.; Zou, J.; Yang, Q. Advancing Few-Shot Named Entity Recognition with Large Language Model. *Appl. Sci.* **2025**, *15*, 3838. [CrossRef]
2. Jiang, J.; Ying, F.; Dhuny, R. Unveiling Technological Evolution with a Patent-Based Dynamic Topic Modeling Framework: A Case Study of Advanced 6G Technologies. *Appl. Sci.* **2025**, *15*, 3783. [CrossRef]
3. Tian, Y.; Guo, X.; Wang, J.; Li, B.; Zhou, S. Video Temporal Grounding with Multi-Model Collaborative Learning. *Appl. Sci.* **2025**, *15*, 3072. [CrossRef]
4. Doumanas, D.; Soularidis, A.; Spiliotopoulos, D.; Vassilakis, C.; Kotis, K. Fine-Tuning Large Language Models for Ontology Engineering: A Comparative Analysis of GPT-4 and Mistral. *Appl. Sci.* **2025**, *15*, 2146. [CrossRef]
5. Cao, X.; Shi, L.; Wang, X.; Duan, Y.; Yang, X.; Zhang, X. MCP: A Named Entity Recognition Method for Shearer Maintenance Based on Multi-Level Clue-Guided Prompt Learning. *Appl. Sci.* **2025**, *15*, 2106. [CrossRef]
6. Yang, J.; Cao, J.; Duan, C. Multi-Task Supervised Alignment Pre-Training for Few-Shot Multimodal Sentiment Analysis. *Appl. Sci.* **2025**, *15*, 2095. [CrossRef]
7. Iaroshev, I.; Pillai, R.; Vaglietti, L.; Hanne, T. Evaluating Retrieval-Augmented Generation Models for Financial Report Question and Answering. *Appl. Sci.* **2024**, *14*, 9318. [CrossRef]
8. Gao, R.; Jiang, L.; Zou, Z.; Li, Y.; Hu, Y. A Graph Convolutional Network Based on Sentiment Support for Aspect-Level Sentiment Analysis. *Appl. Sci.* **2024**, *14*, 2738. [CrossRef]

applied sciences

MDPI

Article

Advancing Few-Shot Named Entity Recognition with Large Language Model

Yuhui Xiao, Jianjian Zou and Qun Yang *

College of Computer Science and Technology, Nanjing University of Aeronautics and Astronautics, Nanjing 211106, China; xiaoyuhui@nuaa.edu.cn (Y.X.); jj.zou@nuaa.edu.cn (J.Z.)
* Correspondence: qun.yang@nuaa.edu.cn

Abstract: Few-shot named entity recognition (NER) involves identifying specific entities using limited data. Metric learning-based methods, which compute token-level similarities between query and support sets to identify target entities, have demonstrated remarkable performance in this task. However, their effectiveness deteriorates when the distribution of the support set differs from that of the query set. To address this issue, we propose a novel approach that leverages the synergy between the large language model (LLM) and the metric learning-based few-shot NER approach. Specifically, we use the LLM to refine low-confidence predictions produced by the metric learning-based few-shot NER model, thus improving overall recognition accuracy. To further reduce the difficulty of entity classification, we introduce multiple label-filtering strategies to reduce the difficulty for LLMs in performing entity classification. Furthermore, we explore the impact of prompt design on enhancing NER performance. Experimental results show that the proposed method increases the micro-F1 score on Few-NERD and CrossNER by 0.86% and 4.9%, respectively, compared to previous state-of-the-art methods.

Keywords: few-shot named entity recognition; large language models; prompt engineering

1. Introduction

Named entity recognition (NER), as a core task in information extraction, aims to identify words or phrases in text that belong to specific semantic categories. Few-shot NER focuses on leveraging limited data to recognize target categories.

In recent years, studies on few-shot NER have predominantly utilized metric learning, which calculates token-level similarities between query and support samples to identify target entities. The core principle of metric learning is to learn distance or similarity functions that ensure favorable geometric properties in the particular representation space. This approach typically comprises two stages: embedding functions and metric-based classification. The embedding function maps high-dimensional vectors into a low-dimensional space, positioning entities of the same type closely together while distinguishing entities of different types. Metric-based classification then assigns labels to entities based on a chosen distance or similarity metric. Das et al. [1] propose CONTaiNER, which optimizes Gaussian distribution embeddings through contrastive learning, allowing the model to acquire more generalized feature representations and demonstrate superior performance in few-shot NER tasks. Similarly, Huang et al. [2] introduce COPNER, which implements a prompt-based contrastive learning mechanism with class-anchor words, facilitating explicit alignment of entities with their corresponding class labels. However, the metric learning paradigm exhibits a significant limitation: its reliance on token-level similarity between query and support sets can lead to substantially degraded recognition performance when their distributions diverge.

Large language models (LLMs) have recently made significant advancements in tasks such as semantic understanding, question answering, and logical reasoning [3–6]. Several studies have evaluated the performance of LLMs on NER tasks, showing that they still lag behind existing specialized methods [7,8]. Furthermore, enhancing the performance of LLMs in NER tasks typically requires dividing them into multiple subtasks, thereby increasing resource consumption and computational costs [9]. An alternative approach entails constructing question–answer pairs to fine-tune LLM [10]. However, this approach escalates computational resource demands and conflicts with the constraints of few-shot learning.

To address these issues, we propose a collaborative approach that combines small and large language models to enhance entity recognition accuracy in few-shot settings. Specifically, we initially deploy the small language model (SLM) fine-tuned with metric learning for few-shot NER. This few-shot NER model then performs preliminary entity identification and classification of the query. When token-level similarity between the query and support sets is low, the query is often distant from feature clusters, resulting in reduced entity prediction accuracy. To address this issue, we utilize an LLM for secondary recognition of queries exhibiting low token-level similarity to the support set. LLMs typically exhibit higher accuracy in these scenarios by drawing on their extensive prior knowledge, contextual understanding, and capacity to incorporate provided entity-type definitions. Thus, our method effectively mitigates classification uncertainty and performance degradation stemming from low cosine similarity and limited semantic coverage in support sets, enhancing the performance of few-shot NER.

To further enhance the performance of LLMs in few-shot NER tasks, we introduce multiple label-filtering strategies. By restricting the label scope predicted by an LLM, we reduce the complexity of entity classification. We also design structured prompts to improve entity classification accuracy. Additionally, we investigate three zero-shot prompting strategies—chain-of-thought prompting [11], metacognitive prompting [12], and self-verification prompting [13]—and evaluate their impact on the performance of few-shot NER.

Overall, the domain-specific knowledge of an SLM complements the generalization capabilities of an LLM, thereby enhancing the performance of a few-shot NER task. Experimental results demonstrate that our collaborative approach, when integrated with various open-source LLMs, improves F1 scores across multiple few-shot NER benchmarks.

Our primary contributions are summarized as follows:

- We propose a hybrid framework combining the metric learning-based few-shot NER model and the LLM for improving the performance of few-shot NER, addressing limitations of metric learning when support set coverage is insufficient;
- We introduce label-filtering strategies, such as retaining top-N labels, to reduce classification complexity and enhance LLM performance;
- We design structured prompts to enhance entity classification accuracy and investigate three zero-shot prompting strategies, evaluating their impact on few-shot NER performance.

2. Related Work

2.1. Few-Shot Named Entity Recognition

Few-shot NER is the task of identifying and classifying named entities in text when only a limited amount of labeled data are available for each entity type. In recent years, metric learning frameworks such as prototypical networks and contrastive learning have seen widespread application in few-shot NER tasks, significantly improving recognition performance in the few-shot setting. For instance, Fritzler et al. [14] first apply prototypical networks to NER by generating prototypes for each entity type and classifying test samples based on their distance to these prototypes. Since the non-entity category ("O") does not need to closely align with any specific prototype in the vector space, the similarity score is adjusted using hyperparameters and optimized through a modified softmax function to improve classification accuracy for non-entity tokens. The proposal by Hou et al. [15] combines prototypical networks with label semantic information. L-TapNet projects both label prototypes and their semantic representations into a feature space that separates different label distributions, thereby enhancing domain adaptability and cross-domain generalization.

Despite the effectiveness of prototypical networks in optimizing distances between prototypes of entity types and samples, they often struggle to learn accurate and discriminative prototypes in the presence of noisy labels. To address this challenge, researchers have introduced contrastive learning as an alternative to traditional prototype-based methods. Contrastive learning simultaneously increases the similarity of positive instances and decreases the similarity of negative instances, thereby encouraging semantically similar entities to cluster while pushing dissimilar entities apart [16]. For example, Das et al. [1] developed CONTaiNER, which optimizes distances in the Gaussian embedding space via contrastive learning to learn more generalized representations, ultimately achieving superior performance in few-shot NER tasks. Xiao et al. [17] propose LACNER, which integrates contrastive learning with label words to enhance token representations and employ dropout-induced noise as a form of data augmentation. However, entity classification performance under metric learning frameworks heavily depends on the quality and distribution of the support set. When the support set lacks certain semantic features, the model's performance significantly declines and becomes more vulnerable to noisy sample interference.

Although metric learning has advanced few-shot NER on the model side, data scarcity remains a challenge. A straightforward solution is to synthesize new data. Current research on word-level transformations for NER primarily focuses on entity replacement or context (non-entity) substitution. For example, Dai et al. [18] explored simple augmentation strategies such as label-wise word replacement, synonym substitution, and random shuffling of words within predefined text segments for few-shot NER. COSINER, a cosine-similarity-based vocabulary replacement strategy, showed significant improvements over random replacement methods [19]. Nevertheless, because NER requires precise syntactic and semantic understanding, direct substitution can disrupt sentence integrity and negatively impact training outcomes.

2.2. Named Entity Recognition Based on LLM

By training on vast amounts of data, LLMs exhibit robust few-shot learning and generalization capabilities [20]. During inference, supplying a few demonstrations enables the LLM to perform tasks accurately without fine-tuning [21]. Leveraging this characteristic, some studies have applied the LLM to NER via prompt engineering. For instance, Xie et al. [7] decompose the NER task into simpler subproblems and extract entities sequentially by type. Wei et al. [9] introduce a two-stage dialogue paradigm: the first stage identifies potential element types (e.g., entities, relationships, events), while the second stage extracts specific mentions of each type. In another study, Wang et al. [22] propose GPT-NER, transforming sequence-labeling-based NER into a text generation task that requires the model to mark entity positions while copying other content verbatim. Although these works offer initial insights into integrating the LLM with NER through in-context learning, they still lag behind state-of-the-art few-shot NER methods.

2.3. Language Model Plug-Ins

In addition to their strong in-context learning abilities, LLMs can be further enhanced by integrating external tools. Several studies have explored this idea by leveraging the SLM to complement LLM capabilities in natural language understanding (NLU) tasks. For instance, Xu et al. [23] suggest using the SLM as a plug-in to improve the LLM's NLU capabilities and mitigate instability in in-context learning. However, their work does not specifically target named entity recognition, nor does it consider a few-shot or metric-learning-based setting.

Ma et al. [24] propose an approach that integrates span-based few-shot NER models [25,26] with LLM-based re-ranking of uncertain entities. Their work primarily focuses on span-based NER, where entity spans are first identified using local models before being refined by the LLM. This approach differs fundamentally from our metric-learning-based framework, where entity recognition is formulated as a similarity computation problem rather than a span-detection task. Similarly, Zhang et al. [27] employ uncertainty estimation to facilitate collaboration between LLM and the local NER model, aiming to reduce errors in recognizing unseen entities. While effective, their framework is designed for span-based or fully supervised settings, rather than the few-shot setting we investigate.

In contrast, our work explores the integration of the SLM and the LLM in a metric learning-based few-shot NER framework, an area that remains underexplored. Metric learning-based NER presents unique challenges, such as sensitivity to distributional shifts between the support and query sets, which span-based or fully supervised models do not explicitly address. Additionally, unlike prior studies that rely heavily on manually designed chain-of-thought (CoT) prompts, our method reduces reliance on handcrafted demonstrations, making it more adaptable to different NER tasks.

3. Methodology

3.1. Method Overview

Figure 1 illustrates our proposed method, which consists of four key steps. First, we utilize the few-shot NER model to identify named entities by computing cosine similarity scores between the query and support sets. Second, words within each entity requiring re-prediction are filtered using a similarity threshold, while restricting the range of visible labels for the large language model to prevent unnecessary label predictions. Third, the filtered candidate words, label scope, type definitions, and few-shot examples are incorporated into a structured prompt template. We apply various prompt engineering techniques to process these inputs, then feed the prompts into the LLM to generate final predictions. Finally, we integrate predictions from both the few-shot NER model and LLM to generate the final entity recognition results.

3.2. Select the Candidate Word

In this study, we choose LACNER as the metric learning-based few-shot NER model [17]. Specifically, given a support set $X_{\text{sup}} = \{x_1^{\text{sup}}, x_2^{\text{sup}}, \ldots, x_n^{\text{sup}}\}$, which consists of multiple sequences, each containing n words, and a query (test) set $X_{\text{test}} = \{x_1^{\text{test}}, x_2^{\text{test}}, \ldots, x_m^{\text{test}}\}$, which also consists of multiple sequences, each containing m words. In the few-shot NER method based on metric learning, X_{sup} and X_{test} are fed into the SLM to generate representation. Then, we compute the cosine similarity between h_i^{test} and h_k^{sup}. The label corresponding to the support sample with the highest similarity is assigned as the entity type of the query. Formally,

$$y_i^{\text{test}} = \underset{(x_k^{\text{sup}}, y_k^{\text{sup}}) \in X_{\text{sup}}}{\arg\max} \; \text{CosineSimilarity}(h_i^{\text{test}}, h_k^{\text{sup}}) \quad (1)$$

where $(x_k^{\text{sup}}, y_k^{\text{sup}})$ denotes the k-th support sample and its label, while h_i^{test} and h_k^{sup} represent the semantic embeddings of the query and support samples, respectively.

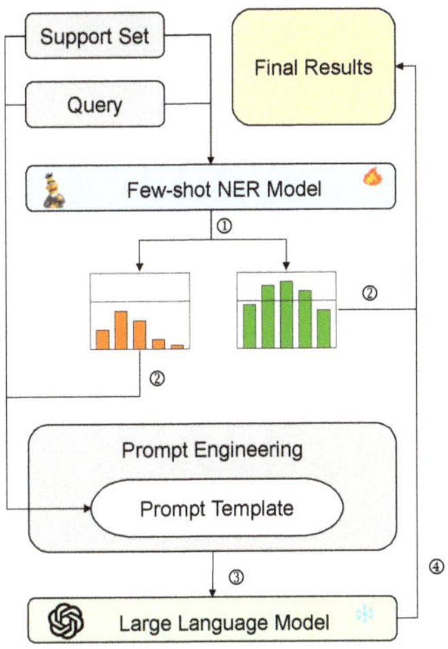

Figure 1. The overview of our method. The different numbers correspond to the different stages described in the Method Overview.

Building on this framework, we define the confidence score of the prediction for query as the maximum cosine similarity between that query and the support sets:

$$\text{conf}(x_i^{\text{test}}) = \max_{(x_k^{\text{sup}}, y_k^{\text{sup}}) \in X_{\text{sup}}} \text{CosineSimilarity}(h_i^{\text{test}}, h_k^{\text{sup}}) \tag{2}$$

Meanwhile, the support sets are ranked in descending order based on their cosine similarity with the query, and the corresponding labels are preserved. Specifically,

$$\mathcal{Y}_{\text{sorted}}(x_i^{\text{test}}) = \{y_{\sigma_i(k)}^{\text{sup}}\}_{k=1}^{|X_{\text{sup}}|} \tag{3}$$

where σ_i represents the descending-order ranking of the support sets x_i^{test} according to $\text{CosineSimilarity}(h_i^{\text{test}}, h_k^{\text{sup}})$.

Then, the tuple of prediction result for the query is defined as follows:

$$z_i = (x_i^{\text{test}}, y_i^{\text{test}}, \text{conf}(x_i^{\text{test}}), \mathcal{Y}_{\text{sorted}}(x_i^{\text{test}})) \tag{4}$$

Finally, we select specific words that require re-prediction. Given a prediction result tuple z_i, if the confidence score of the query is less than or equal to the threshold θ,

$$\text{conf}(x_i^{\text{test}}) \leq \theta \tag{5}$$

and the word is part of an entity, it is retained and passed to the LLM for further refinement.

3.3. Limit the Scope of the Labels to Be Predicted

In prompt construction, a common practice is to concatenate all entity types from the dataset before supplying them to the LLM for prediction. However, study shows that the LLM struggle with fine-grained NER tasks [27]. This limitation arises mainly from the large number of label categories and their fine granularity, making it challenging for the LLM to accurately differentiate labels given limited instructions and examples.

Furthermore, including explanations for all entity type definitions increases both input length and computational cost. To alleviate the LLM's cognitive burden and minimize input length, we propose several strategies to limit the scope of the labels to be predicted. These strategies are designed to enhance the efficiency and accuracy of fine-grained NER tasks by reducing the complexity of label prediction.

(1) Retain Top-N Labels

As shown on the left side of Figure 2, we propose a way to retain only the top N labels with high similarity from few-shot NER predictions. This approach ensures that the most relevant entity types are considered while keeping the size of label set manageable. The steps in this process are outlined as follows:

- First, we collect all label predictions for a given query x_i^{test}.
- These labels are then ranked based on the cosine similarity between the support set and the query.
- The top N labels are selected from this ranked list.
- To ensure completeness, the "None" label is always included if it is not already in the selected list.
- To avoid redundancy, any duplicate labels are eliminated from the final selection.

Formally, we define the label filtering process as follows:

$$\mathcal{Y}_{\text{filtered}}(x_i^{\text{test}}) = \text{unique}\big(\text{Top}_N(\mathcal{Y}_{\text{sorted}}(x_i^{\text{test}})) \cup \{\text{None}\}\big), \tag{6}$$

where $\text{Top}_N(\mathcal{Y}_{\text{sorted}}(x_i^{\text{test}}))$ selects the top N labels from the ranked label set, $\text{unique}(\cdot)$ removes duplicate labels, and the inclusion of "None" ensures that non-entity predictions are always considered.

(2) Retain the Top-N Unique Labels

As illustrated on the right side of Figure 2, this way focuses on ensuring diversity within the selected labels. The primary steps are as follows:

- Start with the ranked label predictions for a given query x_i^{test}.
- Iterate through the ranked labels and select unique ones until N unique labels are collected.
- If the "None" label is not already included, append it to the final set.
- Exclude any duplicate labels from the final selections.

Mathematically, this process is expressed as follows:

$$\mathcal{Y}_{\text{filtered}}(x_i^{\text{test}}) = \{y_k^{\text{sup}} | y_k^{\text{sup}} \in \mathcal{Y}_{\text{sorted}}(x_i^{\text{test}})\}_{k=1}^{N} \cup \{\text{None}\}, \tag{7}$$

where $\left\{y_k^{\text{sup}}\right\}_{k=1}^{N}$ represents the first N unique labels selected from the ranked label set.

(3) Retain All Entity Labels

In this setting, no label filtering is applied, meaning that the model has access to the entire set of possible entity labels. The process can be summarized as follows:

- All entity labels appearing in the dataset are provided to the model during inference.
- There is no restriction on the number of labels.
- The "None" label is included by default to represent cases without named entities.

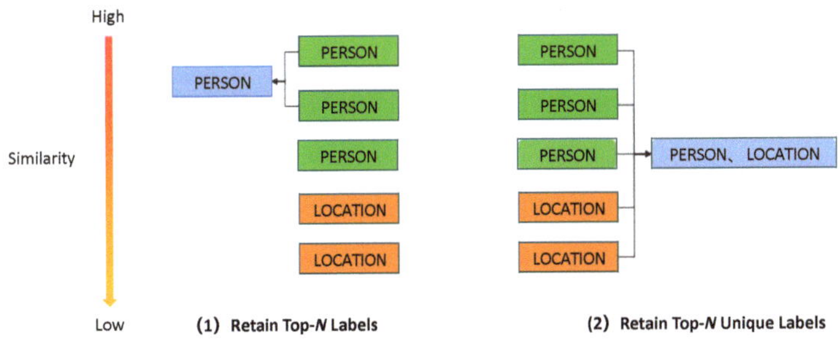

Figure 2. The overview of two ways for limiting the scope of labels to be predicted.

3.4. Prompt Design for LLM

Large language models acquire knowledge from prompts by combining task descriptions with demonstrations, enabling them to identify and perform new tasks without explicit gradient updates. As illustrated in Figure 3, we design a structured prompt for the classification of candidate word. This prompt consists of eleven components: *Context, Objective, Type Definition, Few-shot Examples, Sentence, Types, Candidate, Question, Prompt Engineering Guide, Answer Audience, and Answer Format*. In this section, we focus on three key components: Type Definition, Few-shot Examples, and Prompt Engineering Guide.

(1) Type Definition

Across different datasets, entity types may share the same label but have distinct definitions. Relying solely on the LLM's inherent knowledge may fail to capture these nuances. To mitigate this issue, we create dataset-specific entity-type definitions. Specifically, we compile official annotation guidelines, relevant examples, and entity-type details from the dataset, then utilize GPT-4o to generate descriptive definitions based on this information.

(2) Few-shot Examples

First, we generate sentence embeddings for each sentence in support set using the sentence embedding model. Next, we rank the sentences based on their semantic similarity to the test sample and select the top four most relevant sentences as few-shot examples. Moreover, LLM performance in few-shot learning is sensitive to positional bias, where the order of examples affects the results [28]. To mitigate this effect, examples with higher similarity to the test sample are positioned closer to it in the prompt sequence.

(3) Prompt Engineering Guide

To harness the reasoning capabilities of the LLM, we introduce three zero-shot prompt engineering techniques: COT prompting, metacognitive prompting, and self-verification prompting. COT prompting operates by decomposing complex tasks into a sequence of subproblems, reformulating the conventional "input-output" structure into an "input–reasoning chain–output" framework. We implement the zero-shot COT strategy by incorporating the instruction "Answer the question step by step" into the prompt, incrementally guiding the model's reasoning process.

```
<Context>
You are performing a named entity recognition (NER) task. ...
<Objective>
Your objective is to assign the most appropriate entity type from <Types> to the given <Candidate>
in the <Sentence> according to the <Type Definition>.
<Type Definition>
Here is the refined entity class information for named entity recognition: ...
<Few-shot Examples>
## Example 1
Sentence: Tesla moved to Texas
Answer:
- The entity type of "Tesla" is "organization".
- The entity type of "Texas" is "location".
<Sentence>
Volkswagen was founded in Germany
<Types>
organization, none
<Candidate>
Volkswagen
<Question>
Please refer to the <Type Definition> and select the most relevant type (from <Types>) for
<Candidate> in the <Sentence>.
<Prompt Engineering Guide>
<Answer Audience>
The intended audience for this answer is software developers. It is crucial to follow the <Answer
Format> strictly. Return <Candidate> exactly as it appears, without any modifications or corrections.
<Answer Format>
# The entity type of "<Candidate>" is "<Types>". #
Answer:
```

Figure 3. The template of candidate word classification prompt.

While COT prompting is highly effective for arithmetic and commonsense reasoning, capturing task semantics and implicit context is more critical for NLU tasks. To address this, we introduce metacognitive prompting, which emulates human metacognition to enhance the model's understanding of task requirements. The metacognitive prompt is shown in Table 1. The metacognitive prompting consists of five key steps:

1. Text Interpretation: Guiding the model to understand the sentence and candidate word, including their positions and roles, similar to human understanding.
2. Preliminary Judgment: Conducting an initial classification of the candidate's type based on type definitions and constraints, reflecting human early judgment.
3. Critical Evaluation: Reviewing the preliminary classification. If uncertainty remains, cross-checking type definitions and few-shot examples to refine the classification, akin to human self-reflection.
4. Decision: Confirming the final entity type and offering a concise rationale, paralleling human decision-making.
5. Confidence Assessment: Evaluating confidence in the final output, simulating the human process of judging reliability.

Table 1. Metacognitive prompt.

Sections	Content
Text Interpretation	Understand the given Sentence and the Candidate. Clarify the Candidate's position and role within the Sentence.
Preliminary Judgment	Based on the <Type Definition> and the given Types, make an initial guess about which entity type (if any) the Candidate should belong to.
Critical Evaluation	Critically reassess your preliminary judgment. If you are uncertain, carefully review the Type Definitions and Few-shot Examples to ensure you are selecting the most suitable type. If it does not fit any type, consider "none".
Decision	Confirm your final decision on the Candidate's entity type and provide a brief explanation of why you selected that type.
Confidence Assessment	Evaluate your confidence level (0–100%) in this classification and briefly explain why you have that level of confidence.

Moreover, large language models can exhibit hallucination or over-prediction, particularly in NER tasks, where they often label non-entities as valid entities [8,22]. To address this, we introduce self-verification prompting, which enables the model to validate its own predictions. An illustrative example of self-verification prompt is shown in Figure 4. In practice, we use the multi-turn dialogue format, sequentially providing the candidate classification prompt, prediction results, and self-verification prompt to the LLM. If the verification outcome aligns with the initial prediction, we retain the original result; otherwise, the candidate is classified as a non-entity.

By applying the above prompt design methods, the LLM can refine the classifications and combine these results with those generated by the few-shot NER model to produce the final predictions.

```
You have determined that the `<Candidate>` in the `<Sentence>` might be classified as `organization`
based on the `<Type Definition>`.

<First-stage Classification Answer>
Sentence: Tesla moved to Texas.
Candidate: Tesla Answer: The entity type of "Tesla" is "organization".

<Question>
Please confirm whether `<Candidate>` in the sentence is truly an entity of type `organization` or if it
should instead be `none`.

If it is an entity of type `organization`, answer yes.
If it is not an entity , answer no.
```

Figure 4. Self-verification prompt.

4. Experiments

4.1. Datasets

We selected two widely datasets for few-shot NER tasks: Few-NERD and Cross-NER. Both datasets span multiple domains and entity types, facilitating a comprehensive evaluation of model performance across diverse scenarios.

Few-NERD, introduced by Ding et al. [29], is specifically designed for few-shot NER tasks. It comprises eight coarse-grained and 66 fine-grained entity types, serving as a comprehensive benchmark for evaluating model performance in few-shot settings. Additionally, it defines two tasks with distinct granularities: INTRA and INTER. In the INTRA setting, the training, validation, and test sets contain distinct coarse-grained categories.

In the INTER setting, they may share the same coarse-grained categories, but the fine-grained categories remain mutually exclusive. Few-NERD is distributed under the Creative Commons Attribution-ShareAlike 4.0 International License (CC BY-SA 4.0).

CrossNER comprises NER datasets from four domains—OntoNotes 5.0 [30], CoNLL-03 [31], WNUT-17 [32], and GUM [33]—and serves as a benchmark for evaluating model generalization in cross-domain environments. CrossNER is distributed under the Apache License 2.0.

4.2. Experimental Setup

To reduce experimental costs, we sample subsets from the test datasets. For the Few-NERD, we randomly select 100 support sets and 100 query sets are randomly selected. For the CrossNER, we select 200 support sets and 200 query sets in the 1-shot setting, and 100 support sets and 100 query sets in the 5-shot setting. In the main experiments, chain-of-thought prompting, metacognitive prompting, and self-verification prompting are not applied. Experiments related to prompt engineering are discussed in Section 4.7.

We choose LACNER as the metric learning-based few-shot NER model. In the experiments, LACNER's hyperparameters and setting are exactly the same as those reported by Xiao et al. [17]. For LLMs, we employ various open-source models. To expedite inference, we employ vLLM and set the temperature parameter to 0 to ensure accuracy in the generated results. All models, except for DeepSeek-V2.5, which is accessed via an external API, are executed on a single A100 GPU (Nvidia, Santa Clara, CA, USA).

To ensure fairness, we generate three sets of few-show NER model predictions, each using a different random seed, and subsequently feed these predictions into the LLM to minimize randomness. Each experiment is repeated three times, and the average micro-F1 score is reported. We use BGE-large-en-v1.5 as the sentence embedding model. Furthermore, all F1 scores mentioned in this paper refer to micro-F1.

4.3. Baseline

We select different baseline from two distinct perspectives. First, we adopt the state-of-the-art method LACNER as metric learning-based few-shot NER model. Second, we include several LLMs as baselines, which employ in-context learning to address few-shot NER tasks. Adopting the prompt design methodology outlined in [9], we partition NER into two phases—entity recognition and entity classification—progressively addressing each phase via multi-turn dialogue. Furthermore, LLM Rerank combines the SLM and the LLM for few-shot NER tasks, leveraging the SLM for initial filtering and the LLM for re-ranking [24].

4.4. Analysis of the Impact of Thresholds on Model Performance

By setting threshold θ appropriately, uncertain or potentially misclassified samples are redirected to the LLM for re-prediction. The process of appropriately setting the threshold θ is detailed in Appendix A. To validate the effectiveness of the LLM in handling low-confidence predictions, we conduct experiments on the CoNLL-03 using Qwen-2.5-14B. As shown in Figure 5, we explore two threshold configurations: (1) setting $\theta = -1$, where all samples are classified only by LACNER, and (2) setting $\theta = 1$, where the LLM re-predicts every entity word identified by LACNER.

Figure 5 demonstrates that for low-confidence predictions, LACNER $_{w/\ Qwen2.5-14B}$ achieves the higher F1 score than LACNER alone, whereas for predictions with higher confidence, LACNER alone performs better. It shows that when the confidence of prediction is low, the LLM leverages its extensive world knowledge and reasoning capabilities to correct recognition errors. Thus, integrating the LLM with the metric-based few-show NER model enhances overall performance.

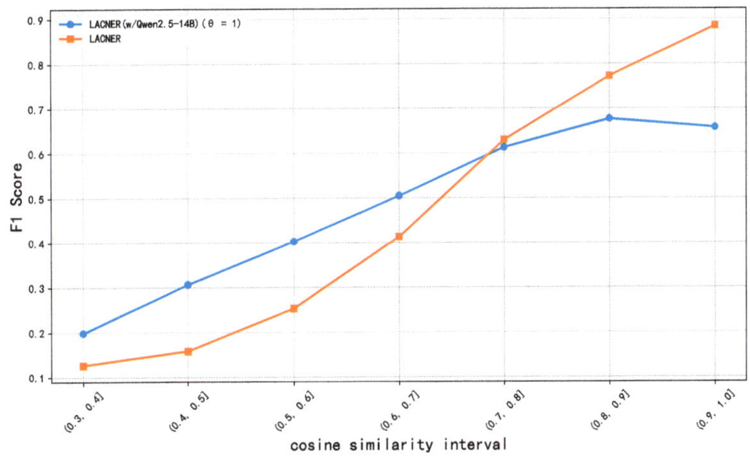

Figure 5. Comparison of micro-F1 of models with different cosine similarity intervals.

We next analyze the impact of threshold selection on the performance of model. As shown in Figure 6, we evaluate LACNER $_{w/\ Qwen2.5-14B}$ by choosing ten threshold points at intervals of 0.1. The results show that as θ varies, the F1 score initially rises before declining. At lower thresholds, the LLM corrects many misclassified samples, leading to improved overall performance. However, at higher thresholds, the LLM erroneously modifies correct predictions, resulting in performance degradation. This trend suggests that applying the LLM to high-confidence samples may introduce additional errors, diminishing overall performance. Thus, selecting an optimal threshold is crucial for maximizing the collaborative benefits of both the few-shot NER model and LLM.

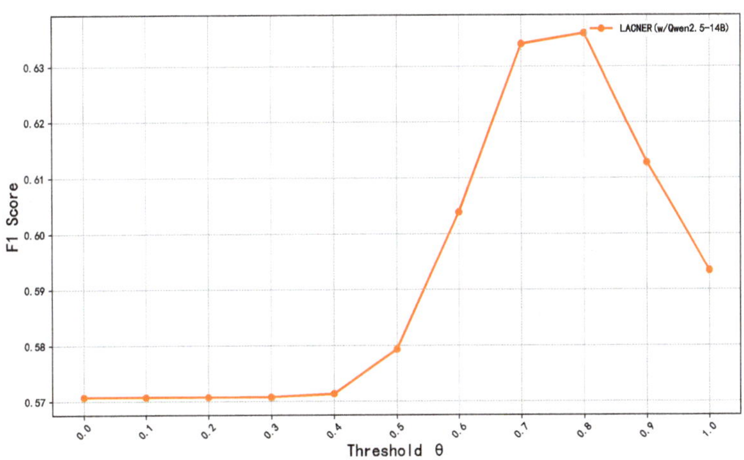

Figure 6. Comparison of performance on CoNLL across various thresholds using the micro-F1 metric (%).

4.5. Comparison with Other Methods

In this section, we conduct experiments on various open-source LLMs with different parameters, comparing the performance of different methods and exploring the contributions of various LLM to LACNER. The experimental results are shown in Tables 2 and 3.

Table 2. Comparison of performance on Cross-NER with the micro-F1 metric (%).

Models	1-Shot					5-Shot				
	CONLL	GUM	WNUT	Onto	Avg.	CONLL	GUM	WNUT	Onto	Avg.
Qwen2.5-7B	51.59	18.82	33.84	37.63	35.47	51.62	18.35	35.31	35.31	35.15
Llama3.1-8B	45.90	18.62	29.51	26.97	30.25	42.76	16.52	25.07	25.07	27.36
Gemma2-9B	54.22	20.54	39.17	39.69	38.41	53.80	20.28	34.31	34.31	35.68
Qwen2.5-14B	62.43	20.29	45.01	44.28	43.0	61.87	21.09	37.85	41.46	40.57
DeepSeek-V2.5	62.71	24.96	48.65	51.87	47.05	63.17	27.80	44.33	49.95	46.31
LLM Rerank	58.62	22.65	46.73	51.16	44.79	70.85	40.12	53.55	64.69	57.30
LACNER	57.08	21.65	40.62	46.63	41.50	70.50	40.12	48.92	63.61	55.79
LACNER $_{w/\ Qwen2.5-7B}$	59.80	23.71	46.90	51.19	45.40	71.51	39.80	53.56	65.48	57.59
LACNER $_{w/\ Llama3.1-8B}$	58.61	23.43	44.48	50.74	44.32	71.31	39.58	51.13	65.08	56.78
LACNER $_{w/\ Gemma2-9B}$	60.36	23.69	48.19	53.12	46.34	72.48	39.82	53.09	65.33	57.68
LACNER $_{w/\ Qwen2.5-14B}$	63.90	23.92	49.61	55.49	48.23	**72.78**	40.48	55.03	67.12	58.85
LACNER $_{w/\ DeepSeek-V2.5}$	**64.18**	**25.25**	**50.36**	**57.42**	**49.30**	72.75	**40.49**	**55.83**	**67.56**	**59.16**

Best results are indicated in bold.

As shown in Table 2, the performance of few-shot NER methods based on the LLM improves significantly with increasing the parameter of the LLM. For example, in the 1-shot setting, when the parameters of the Qwen2.5 series increase from 7B to 14B, the F1 score on all datasets improves. Specifically, Qwen2.5-14B achieves up to 10.84% higher F1 score compared to Qwen2.5-7B. Additionally, DeepSeek-V2.5, which has the largest number of parameters, exhibits the maximum F1 improvement of 24.9% over Llama3.1-8B. It indicates that the larger parameter LLM possess richer world knowledge and stronger reasoning capabilities, leading to better generalization and few-shot learning performance in extremely low-resources.

However, as the number of examples increases, the performance gains of the LLM tend to plateau, meaning the benefit from more examples becomes limited. In contrast, LACNER leverages metric learning to fully capture domain-specific knowledge, and its performance gap with the LLM widens as the number of training samples grows.

Re-identifying low-confidence entity words filtered by LACNER using large models leads to significant improvements across all settings. In the 1-shot setting, compared to using LACNER alone, LACNER $_{w/\ Qwen2.5-14B}$ achieves the F1 improvement ranging from 2.27% to 8.99%, with an average gain of 6.73%. These results indicate that the extensive knowledge and strong reasoning capabilities of the LLM compensate for LACNER's limitations in representation learning for certain entity types, as well as classification errors due to limited support set examples, thereby enhancing overall performance. In the 5-shot setting, as LACNER learns entity representations more effectively, fewer samples require re-identification by the LLM, leading to smaller performance gains compared to the 1-shot setting. Nevertheless, LACNER $_{w/\ Qwen2.5-14B}$ achieves the average F1 improvement of 3.06%.

Overall, the integration of the LLM enhances LACNER's performance; both Qwen2.5-14B and DeepSeek-V2.5 significantly boost LACNER's performance in the 1-shot and 5-shot settings. These findings confirm that the parameter scale and pretraining quality of the LLM significantly impact their generalization and knowledge transfer capabilities in downstream tasks.

Furthermore, compared with LLM Rerank, LACNER $_{w/\ Qwen2.5-14B}$ improves the average F1 score by 3.44% in the 1-shot setting and by 1.55% in the 5-shot setting. It indicates that restricting the scope of predicted labels can enhance the classification accuracy of the LLM. Additionally, the structured prompting approach proposed in this study outperforms the multiple choice prompting used in LLM Rerank by improving the LLM's understanding of task requirements and providing relevant background knowledge.

Table 3. Comparison of performance on Few-NERD with the micro-F1 metric (%).

Models	INTRA					INTER				
	1∼2shot		5∼10shot		Avg.	1∼2shot		5∼10shot		Avg.
	5way	10way	5way	10way		5way	10way	5way	10way	
Qwen2.5-7B	21.76	22.54	26.47	24.75	23.88	28.51	27.88	35.18	33.87	31.36
Llama3.1-8B	18.59	19.96	23.52	22.62	21.17	20.98	19.47	25.77	25.42	22.91
Gemma2-9B	26.38	26.47	28.11	26.98	26.99	30.51	31.09	33.52	34.85	32.49
Qwen2.5-14B	36.42	38.0	42.15	39.63	39.05	48.15	49.80	50.40	50.63	49.75
DeepSeek-V2.5	39.06	37.56	44.73	44.64	41.50	47.88	46.80	52.0	52.94	49.91
LLM Rerank	62.77	60.15	74.14	70.84	66.98	68.0	68.64	73.40	75.79	71.46
LACNER	61.92	58.11	74.0	70.58	66.15	66.98	67.28	72.87	75.73	70.72
LACNER $_{w/\ Qwen2.5-7B}$	63.71	59.21	75.50	71.15	67.39	**69.79**	69.44	73.60	76.49	**72.33**
LACNER $_{w/\ Llama3.1-8B}$	63.56	58.38	75.10	70.86	66.98	68.30	68.11	72.82	75.41	71.16
LACNER $_{w/\ Gemma2-9B}$	**64.83**	59.37	75.46	71.19	67.71	68.51	68.64	73.26	75.73	71.54
LACNER $_{w/\ Qwen2.5-14B}$	64.01	60.12	**75.69**	71.80	67.91	69.48	69.06	**73.86**	**76.54**	72.24
LACNER $_{w/\ DeepSeek-V2.5}$	**64.83**	**60.50**	75.65	**71.81**	**68.20**	69.27	**69.46**	73.67	76.36	72.19

Best results are indicated in bold.

Compared with CrossNER, the Few-NERD contains a greater variety of entity labels with finer granularity, a larger labeling space, and higher differentiation complexity. On CrossNER, the LLM performed well in the 1-shot setting, benefiting from their extensive prior knowledge. However, when applied independently to Few-NERD, the LLM demonstrated lower performance, particularly in the INTRA, where the disparity is more pronounced. It suggests that as task complexity and label granularity increase, LLMs face greater challenges.

Although integrating an LLM with LACNER continues to enhance performance, the improvements become increasingly nuanced. Specifically, in the INTRA setting, LACNER $_{w/\ Qwen2.5-14B}$ improves the average F1 score by 1.76% compared to LACNER, while in the INTER setting, it achieves a 1.52% improvement. The disparity in these collaborative improvements narrows, suggesting that as task complexity increases, LLMs face greater challenges.

Finally, compared with LLM Rerank, LACNER $_{w/\ Qwen2.5-14B}$ achieves the average F1 improvement of 0.95% in the INTRA setting and 0.78% in the INTER setting. Moreover, unlike LLM Rerank, it removes the need for manually creating the COT examples.

4.6. Analysis of the Impact of Predicted Label Scope on Collaborative Capabilities

As shown in Table 4, the advantage of Qwen2.5-14B over LACNER gradually decreases as the number of entity types increases. In some cases, LACNER even outperforms Qwen2.5-14B. As the number of entity types grows, distinguishing fine-grained categories with subtle differences becomes more challenging. In contrast, the metric learning-based few-shot NER model, trained on diverse examples, effectively learns to differentiate these subtle nuances. Thus, implementing an appropriate label scope filtering strategy during collaboration is crucial for reducing confusion in the LLM.

Table 4. Performance comparison of Qwen2.5-14B and LACNER across datasets with varying numbers of entity types in the 1-shot setting.

	CONLL	**GUM**	**WNUT**	**OntoNotes**	**Few-NERD**
Number of entity types	4	12	7	18	17
Micro-F1(Qwen2.5-14B)	62.43	20.29	45.01	44.28	25.17
Micro-F1(LACNER)	57.08	21.65	40.62	46.63	63.57
Δ Micro-F1 (Qwen2.5-14B, LACNER)	5.35	−1.36	4.39	−2.35	−38.40

Tables 5 and 6 present the impact of varying the scope of predicted labels on the Few-NERD and CrossNER. Table 5 shows that when all entity types are provided, the classification space for the LLM becomes excessively large. Compared to the "retain top-1 label" strategy, performance drops by up to 1.72%, suggesting that limiting the label scope improves model focus. Similarly, Table 6 reveals a comparable trend in the 5-shot setting of CrossNER, where the "retain top-N labels" strategy achieves the best performance. This approach prevents the LLM from being overwhelmed by excessive predicted labels while enabling it to reference additional labels when the metric learning-based few-shot NER model's predictions are uncertain or when test samples lie near cluster boundaries. However, in the 1-shot setting of CrossNER, the significant domain discrepancy between the source and target domains leads to suboptimal representations learned by LACNER. In this case, providing all predicted labels more effectively leverages the LLM's capabilities.

Table 5. Effects of predicted label scope on the Few-NERD with the micro-F1 metric (%).

Scope of the Labels to Be Predicted	**INTRA**					**INTER**				
	1∼2shot		**5∼10shot**		**Avg.**	**1∼2shot**		**5∼10shot**		**Avg.**
	5way	**10way**	**5way**	**10way**		**5way**	**10way**	**5way**	**10way**	
All	62.29	59.13	74.58	71.10	66.78	67.73	67.75	73.31	75.70	71.12
Top 1 Labels	**64.01**	**60.12**	**75.69**	71.76	**67.90**	**69.48**	**69.06**	73.83	**76.54**	**72.23**
Top 2 Labels	64.0	60.10	75.45	71.77	67.83	69.31	68.94	**73.86**	76.44	72.14
Top 2 Unique Labels	63.94	59.95	75.38	71.76	67.76	69.39	68.59	73.79	76.38	72.04
Top 3 Labels	63.99	60.05	75.45	**71.80**	67.82	69.43	68.61	73.84	76.43	72.08
Top 3 Unique Labels	63.67	59.95	75.28	71.69	67.65	69.50	68.71	73.82	76.39	72.11

Best results are indicated in bold.

Table 6. Effects of predicted label scope on the CrossNER with the micro-F1 metric (%).

Scope of the Labels to Be Predicted	**1-Shot**					**5-Shot**				
	CONLL	GUM	WNUT	Onto	Avg.	CONLL	GUM	WNUT	Onto	Avg.
All	63.76	**23.92**	48.81	**55.49**	**48.0**	**72.78**	40.09	54.41	66.34	58.41
Top 1 Labels	63.53	23.13	49.5	52.94	47.28	72.61	**40.48**	**55.03**	**67.12**	**58.81**
Top 2 Labels	63.78	23.27	**49.61**	53.29	47.49	72.50	40.41	54.88	67.04	58.71
Top 3 Labels	63.65	23.20	49.51	54.41	47.69	72.24	40.30	54.92	67.08	58.64
Top 3 Labels	**63.90**	23.16	49.49	53.42	47.49	72.62	40.48	54.85	67.02	58.74
Top 3 Unique Labels	63.48	23.02	49.06	55.24	47.70	72.30	40.16	54.88	66.87	58.55

Best results are indicated in bold.

4.7. Analysis of the Effectiveness of Various Methods in Prompt Design

In this section, we first examine the effectiveness of incorporating type definitions and few-shot examples in our prompt. We then investigate whether prompt engineering can further enhance overall performance.

(1) Analysis of the Effectiveness of Type Definitions and Few-Shot Examples in In-Context Learning

Table 7 presents the performance of two strategies—type definitions and few-shot examples—using ablation on the CrossNER. The results show that employing both type definitions and few-shot examples yields the best performance under both 1-shot and 5-shot settings. Specifically, in the 1-shot setting, the combined approach outperforms the baseline (which does not apply these strategies) by 1.8%, while in the 5-shot setting, it achieves the 0.52% improvement. These findings suggest that incorporating both type definitions and demonstrations enhances the model's comprehension of the task and entity type definitions, thereby improving classification accuracy. However, on the OntoNotes, removing the demonstrations leads to better performance, indicating that noise in these demonstrations may hinder model understanding. When noise exists in the support set, the classification performance of large and small language models is affected to varying degrees. Large language models, which leverage extensive world knowledge, exhibit stronger robustness against noise, whereas smaller models are more prone to noise interference in few-shot setting.

Table 7. Effect of type definitions and demonstrations on the CrossNER with the micro-F1 metric (%). TD stands for type definition, and FSE stands for few-shot examples.

TD	FSE	1-Shot					5-Shot				
		CONLL	GUM	WNUT	Onto	Avg.	CONLL	GUM	WNUT	Onto	Avg.
✓	✓	**63.90**	**23.92**	**49.61**	55.49	48.23	**72.78**	**40.48**	**55.03**	67.12	**58.85**
×	✓	63.80	23.78	48.88	53.20	47.42	72.77	40.37	54.68	66.82	58.66
✓	×	63.79	23.52	49.51	**56.39**	**48.30**	72.15	40.43	54.78	**67.77**	58.78
×	×	62.80	23.17	48.45	51.29	46.43	71.82	40.35	54.57	66.58	58.33

Best results are indicated in bold; ✓ means this component is used, and × means it is not used.

(2) Analysis of the Effectiveness of Prompt Engineering

Prompt engineering techniques can further stimulate the model's reasoning abilities. We evaluate how different zero-shot prompting approaches affect the performance of model.

As shown in Table 8, metacognitive prompting improves performance by 0.76% in the 1-shot setting and 0.23% in the 5-shot setting compared to our prompt. Emulating human metacognition enhances the model's ability to comprehend and execute NER tasks more effectively. However, zero-shot CoT and self-verification prompting generally lead to performance degradation. The CoT prompting is originally designed for tasks requiring explicit step-by-step reasoning, such as mathematical or logical problems. In contrast, NER primarily relies on textual semantics and contextual understanding, which do not align perfectly with CoT's strengths. For self-verification prompting, if the initial prediction is highly uncertain, the mechanism allows the model to reassess and correct errors, explaining its effectiveness on certain datasets (e.g., CoNLL). However, if the model is overly confident in an incorrect prediction, the verification process may reinforce the error by drawing upon the same underlying knowledge base and reasoning patterns. Therefore, for tasks heavily reliant on semantic understanding, such as NER, prompt designs should prioritize enhancing text comprehension.

Table 8. Impact of different prompt engineering methods on the CrossNER with the micro-F1 metric (%).

Prompt Engineering	1-Shot					5-Shot				
	CONLL	GUM	WNUT	Onto	Avg.	CONLL	GUM	WNUT	Onto	Avg.
Ours	63.90	23.92	49.61	55.49	48.23	72.78	40.48	55.03	67.12	58.85
+ Metacognitive Prompting	64.28	**24.32**	**50.91**	**56.46**	**48.99**	**72.91**	**40.59**	**55.35**	**67.45**	**59.08**
+ Chain of Thought Prompting	64.20	23.99	50.19	55.41	48.45	72.70	40.48	54.92	67.22	58.83
+ Self-Verification Prompting	**64.40**	23.92	49.51	56.33	48.54	72.67	40.42	55.02	66.83	58.74

Best results are indicated in bold.

5. Limitations of the Study

Our proposed method achieves significant improvements in few-shot NER by leveraging the collaboration between small and large language models. However, several limitations warrant consideration for future refinement.

To begin with, our approach uses cosine similarity as the confidence score to determine which predictions from the small language model require refinement by the large language model. Although cosine similarity is effective due to its scale invariance and standardized range—making it a practical choice for metric learning frameworks—it may not fully capture the uncertainty inherent in predictions. For instance, when the support set poorly represents the query set's distribution, cosine similarity might overestimate confidence in erroneous predictions or fail to reflect subtle uncertainties. Alternatives such as Monte Carlo Dropout [34] or Evidential Neural Networks [35], could provide richer uncertainty estimates by modeling prediction variability or distributional properties. Nevertheless, these methods require architectural or training modifications, which are beyond the scope of this study. Investigating such alternatives in future work could enhance the robustness of confidence estimation.

Moreover, the threshold θ, which determines which predictions are refined by the LLM, is manually selected based on the performance of validation. While this approach proved effective in our experiments, it is time-consuming and may not generalize across datasets or domains. Therefore, an adaptive or automated threshold-setting mechanism could improve the method's scalability and practicality.

6. Conclusions

In this paper, we propose a few-shot NER method based on the collaboration of large and small language models. The method uses the SLM to make preliminary predictions on test samples, filtering out entities with lower confidence levels, which are then re-identified and corrected by the large language model. This paper also designs multiple prediction label filtering strategies to restrict the range of labels predicted by the LLM, reducing the difficulty of entity classification. Additionally, we adopt the structured approach to designing prompts, further optimizing experimental results through prompt engineering.

Experiments conducted on the Few-NERD and CrossNER demonstrate that the proposed method enhances the performance of few-shot NER. In both 1-shot and 5-shot settings, the collaborative approach outperforms baseline models, and its versatility has been validated across various open-source LLMs. Ablation studies further confirm the effectiveness of key components such as the candidate word selecting module, prediction label filtering module, structured prompts, and prompt engineering.

Author Contributions: Conceptualization, Y.X. and Q.Y.; methodology, Y.X.; validation, Y.X., J.Z., and Q.Y.; formal analysis, Y.X.; investigation, Y.X.; resources, Q.Y.; data curation, Y.X.; writing—original draft preparation, Y.X.; writing—review and editing, Q.Y.; supervision, Q.Y.; project administration, Q.Y.; funding acquisition, Q.Y. All authors have read and agreed to the published version of the manuscript.

Funding: This research received no external funding.

Institutional Review Board Statement: Not applicable.

Informed Consent Statement: Not applicable.

Data Availability Statement: The dataset we used is the newest Few-NERD Arxiv V6 Version. It can be downloaded at https://ningding97.github.io/fewnerd/ (accessed on 10 February 2025). CrossNER can be downloaded at https://atmahou.github.io/attachments/ACL2020data.zip (accessed on 10 February 2025).

Conflicts of Interest: The authors declare no conflicts of interest.

Abbreviations

The following abbreviations are used in this manuscript:

NER	Named Entity Recognition
LLM	Large Language Model
SLM	Small Language Model
NLU	Natural Language Understanding
COT	Chain-of-Thought

Appendix A

To establish the optimal threshold value θ for each setting, we perform hyperparameter tuning using the validation set. Specifically, we employ a grid search approach, testing θ values in the range of 0.5 to 0.8 with an increment of 0.05. For each candidate θ, we evaluate the model's performance using the micro-F1 score. The resulting optimal θ values for each setting appear in Tables A1 and A2 below. These values then apply in our experiments to achieve the results reported in Section 4.

Table A1. Threshold settings on the CrossNER.

1-Shot				5-Shot			
CONLL	GUM	WNUT	Onto	CONLL	GUM	WNUT	Onto
0.75	0.60	0.75	0.75	0.75	0.55	0.75	0.75

Table A2. Threshold settings on the Few-NERD.

INTRA				INTER			
1∼2shot		5∼10shot		1∼2shot		5∼10shot	
5way	10way	5way	10way	5way	10way	5way	10way
0.50	0.55	0.65	0.65	0.55	0.60	0.60	0.70

References

1. Das, S.S.S.; Katiyar, A.; Passonneau, R.J.; Zhang, R. CONTaiNER: Few-Shot Named Entity Recognition via Contrastive Learning. In Proceedings of the 60th Annual Meeting of the Association for Computational Linguistics (Volume 1: Long Papers), Dublin, Ireland, 22–27 May 2022; pp. 6338–6353. [CrossRef]
2. Huang, Y.; He, K.; Wang, Y.; Zhang, X.; Gong, T.; Mao, R.; Li, C. Copner: Contrastive learning with prompt guiding for few-shot named entity recognition. In Proceedings of the 29th International Conference on Computational Linguistics, Gyeongju, Republic of Korea, 12–17 October 2022; pp. 2515–2527.

3. Zhang, C.; Wu, S.; Zhang, H.; Xu, T.; Gao, Y.; Hu, Y.; Chen, E. Notellm: A retrievable large language model for note recommendation. In Proceedings of the Companion Proceedings of the ACM Web Conference 2024, Singapore, 13–17 May 2024; pp. 170–179. [CrossRef]
4. Jeong, S.; Baek, J.; Cho, S.; Hwang, S.J.; Park, J. Adaptive-RAG: Learning to Adapt Retrieval-Augmented Large Language Models through Question Complexity. In Proceedings of the 2024 Conference of the North American Chapter of the Association for Computational Linguistics: Human Language Technologies (Volume 1: Long Papers), Mexico City, Mexico, 16–21 June 2024; Duh, K., Gomez, H., Bethard, S., Eds.; pp. 7036–7050. [CrossRef]
5. Raina, V.; Gales, M. Question-Based Retrieval using Atomic Units for Enterprise RAG. In Proceedings of the Seventh Fact Extraction and VERification Workshop (FEVER), Miami, FL, USA, 15 November 2024; Schlichtkrull, M., Chen, Y., Whitehouse, C., Deng, Z., Akhtar, M., Aly, R., Guo, Z., Christodoulopoulos, C., Cocarascu, O., Mittal, A., et al., Eds.; pp. 219–233. [CrossRef]
6. Pan, L.; Albalak, A.; Wang, X.; Wang, W. Logic-LM: Empowering Large Language Models with Symbolic Solvers for Faithful Logical Reasoning. In Proceedings of the Findings of the Association for Computational Linguistics: EMNLP 2023, Singapore, 6–10 December 2023; Bouamor, H., Pino, J., Bali, K., Eds.; pp. 3806–3824. [CrossRef]
7. Xie, T.; Li, Q.; Zhang, J.; Zhang, Y.; Liu, Z.; Wang, H. Empirical Study of Zero-Shot NER with ChatGPT. In Proceedings of the 2023 Conference on Empirical Methods in Natural Language Processing, Singapore, 6–10 December 2023; pp. 7935–7956.
8. Han, R.; Yang, C.; Peng, T.; Tiwari, P.; Wan, X.; Liu, L.; Wang, B. An Empirical Study on Information Extraction using Large Language Models. *arXiv* **2023**, arXiv:2305.14450. [CrossRef]
9. Wei, X.; Cui, X.; Cheng, N.; Wang, X.; Zhang, X.; Huang, S.; Xie, P.; Xu, J.; Chen, Y.; Zhang, M.; et al. Chatie: Zero-shot information extraction via chatting with chatgpt. *arXiv* **2023**, arXiv:2302.10205.
10. Zhou, W.; Zhang, S.; Gu, Y.; Chen, M.; Poon, H. Universalner: Targeted distillation from large language models for open named entity recognition. *arXiv* **2023**, arXiv:2308.03279.
11. Kojima, T.; Gu, S.S.; Reid, M.; Matsuo, Y.; Iwasawa, Y. Large language models are zero-shot reasoners. *Adv. Neural Inf. Process. Syst.* **2022**, *35*, 22199–22213.
12. Wang, Y.; Zhao, Y. Metacognitive Prompting Improves Understanding in Large Language Models. In Proceedings of the 2024 Conference of the North American Chapter of the Association for Computational Linguistics: Human Language Technologies (Volume 1: Long Papers), Mexico City, Mexico, 16–21 June 2024; pp. 1914–1926. [CrossRef]
13. Kadavath, S.; Conerly, T.; Askell, A.; Henighan, T.; Drain, D.; Perez, E.; Schiefer, N.; Hatfield-Dodds, Z.; DasSarma, N.; Tran-Johnson, E.; et al. Language models (mostly) know what they know. *arXiv* **2022**, arXiv:2207.05221.
14. Fritzler, A.; Logacheva, V.; Kretov, M. Few-shot classification in named entity recognition task. In Proceedings of the 34th ACM/SIGAPP Symposium on Applied Computing, Limassol, Cyprus, 8–12 April 2019; pp. 993–1000. [CrossRef]
15. Hou, Y.; Che, W.; Lai, Y.; Zhou, Z.; Liu, Y.; Liu, H.; Liu, T. Few-shot Slot Tagging with Collapsed Dependency Transfer and Label-enhanced Task-adaptive Projection Network. In Proceedings of the 58th Annual Meeting of the Association for Computational Linguistics, Online, 5–10 July 2020; pp. 1381–1393. [CrossRef]
16. Khosla, P.; Teterwak, P.; Wang, C.; Sarna, A.; Tian, Y.; Isola, P.; Maschinot, A.; Liu, C.; Krishnan, D. Supervised contrastive learning. *Adv. Neural Inf. Process. Syst.* **2020**, *33*, 18661–18673.
17. Xiao, Y.; Yang, Q.; Zou, J.; Zhou, S. Lacner: Enhancing few-shot named entity recognition with label words and contrastive learning. In Proceedings of the 2024 International Joint Conference on Neural Networks (IJCNN), Yokohama, Japan, 30 June–5 July 2024; pp. 1–8. [CrossRef]
18. Dai, X.; Adel, H. An Analysis of Simple Data Augmentation for Named Entity Recognition. In Proceedings of the 28th International Conference on Computational Linguistics, Barcelona, Spain, 8–13 December 2020; pp. 3861–3867. [CrossRef]
19. Bartolini, I.; Moscato, V.; Postiglione, M.; Sperlì, G.; Vignali, A. COSINER: Context similarity data augmentation for named entity recognition. In *International Conference on Similarity Search and Applications*; Springer: Berlin/Heidelberg, Germany, 2022; pp. 11–24. [CrossRef]
20. Lyu, Y.; Li, Z.; Niu, S.; Xiong, F.; Tang, B.; Wang, W.; Wu, H.; Liu, H.; Xu, T.; Chen, E. Crud-rag: A comprehensive chinese benchmark for retrieval-augmented generation of large language models. *ACM Trans. Inf. Syst.* **2025**, *43*, 1–32. [CrossRef]
21. Brown, T.; Mann, B.; Ryder, N.; Subbiah, M.; Kaplan, J.D.; Dhariwal, P.; Neelakantan, A.; Shyam, P.; Sastry, G.; Askell, A.; et al. Language models are few-shot learners. *Adv. Neural Inf. Process. Syst.* **2020**, *33*, 1877–1901.
22. Wang, S.; Sun, X.; Li, X.; Ouyang, R.; Wu, F.; Zhang, T.; Li, J.; Wang, G. Gpt-ner: Named entity recognition via large language models. *arXiv* **2023**, arXiv:2304.10428.
23. Xu, C.; Xu, Y.; Wang, S.; Liu, Y.; Zhu, C.; McAuley, J. Small Models are Valuable Plug-ins for Large Language Models. In Proceedings of the Findings of the Association for Computational Linguistics ACL 2024, Bangkok, Thailand, 11–16 August 2024; pp. 283–294. [CrossRef]
24. Ma, Y.; Cao, Y.; Hong, Y.; Sun, A. Large Language Model Is Not a Good Few-shot Information Extractor, but a Good Reranker for Hard Samples! In Proceedings of the Findings of the Association for Computational Linguistics: EMNLP 2023, Singapore, 6–10 December 2023; pp. 10572–10601. [CrossRef]

25. Fu, J.; Huang, X.J.; Liu, P. SpanNER: Named Entity Re-/Recognition as Span Prediction. In Proceedings of the 59th Annual Meeting of the Association for Computational Linguistics and the 11th International Joint Conference on Natural Language Processing (Volume 1: Long Papers), Virtual, 1–6 August 2021; pp. 7183–7195. [CrossRef]
26. Chen, W.; Zhao, L.; Luo, P.; Xu, T.; Zheng, Y.; Chen, E. Heproto: A hierarchical enhancing protonet based on multi-task learning for few-shot named entity recognition. In Proceedings of the 32nd ACM International Conference on Information and Knowledge Management, Birmingham, UK, 21–25 October 2023; pp. 296–305. [CrossRef]
27. Zhang, Z.; Zhao, Y.; Gao, H.; Hu, M. Linkner: Linking local named entity recognition models to large language models using uncertainty. In Proceedings of the ACM on Web Conference 2024, Virtual, 13–17 May 2024; pp. 4047–4058. [CrossRef]
28. Zhao, Z.; Wallace, E.; Feng, S.; Klein, D.; Singh, S. Calibrate before use: Improving few-shot performance of language models. In Proceedings of the 38th International Conference on Machine Learning: ICML 2021, Virtual, 18–24 July 2021 ; pp. 12697–12706.
29. Ding, N.; Xu, G.; Chen, Y.; Wang, X.; Han, X.; Xie, P.; Zheng, H.; Liu, Z. Few-NERD: A Few-shot Named Entity Recognition Dataset. In Proceedings of the 59th Annual Meeting of the Association for Computational Linguistics and the 11th International Joint Conference on Natural Language Processing (Volume 1: Long Papers), Virtual, 1–6 August 2021; pp. 3198–3213. [CrossRef]
30. Pradhan, S.; Moschitti, A.; Xue, N.; Ng, H.T.; Björkelund, A.; Uryupina, O.; Zhang, Y.; Zhong, Z. Towards robust linguistic analysis using ontonotes. In Proceedings of the Seventeenth Conference on Computational Natural Language Learning, Sofia, Bulgaria, 8–9 August 2013; pp. 143–152.
31. Sang, E.T.K.; De Meulder, F. Introduction to the CoNLL-2003 Shared Task: Language-Independent Named Entity Recognition. In Proceedings of the Seventh Conference on Natural Language Learning at HLT-NAACL 2003, Edmonton, Canada, 1 May–1 June 2003; pp. 142–147.
32. Derczynski, L.; Nichols, E.; Van Erp, M.; Limsopatham, N. Results of the WNUT2017 shared task on novel and emerging entity recognition. In Proceedings of the 3rd Workshop on Noisy User-Generated Text, Copenhagen, Denmark, 7 September 2017; pp. 140–147. [CrossRef]
33. Zeldes, A. The GUM corpus: Creating multilayer resources in the classroom. *Lang. Resour. Eval.* **2017**, *51*, 581–612. [CrossRef]
34. Gal, Y.; Ghahramani, Z. Dropout as a bayesian approximation: Representing model uncertainty in deep learning. In Proceedings of the 33nd International Conference on Machine Learning: ICML 2016, New York City, NY, USA, 19–24 June 2016; pp. 1050–1059.
35. Zhang, Z.; Hu, M.; Zhao, S.; Huang, M.; Wang, H.; Liu, L.; Zhang, Z.; Liu, Z.; Wu, B. E-NER: Evidential Deep Learning for Trustworthy Named Entity Recognition. In Proceedings of the Findings of the Association for Computational Linguistics: ACL 2023, Toronto, ON, Canada, 9–14 July 2023; pp. 1619–1634. [CrossRef]

applied sciences

MDPI

Article

Unveiling Technological Evolution with a Patent-Based Dynamic Topic Modeling Framework: A Case Study of Advanced 6G Technologies

Jieru Jiang [1], Fangli Ying [2,*] and Riyad Dhuny [3]

[1] Institute of Scientific and Technical Information of Shanghai, Shanghai 200031, China; jrjiang@libnet.sh.cn
[2] Department of Computer Science, State Key Laboratory of Bioreactor Engineering, East China University of Science and Technology, Shanghai 200237, China
[3] Department of Creative Arts, Film and Media Technologies, University of Technology, Mauritius, La Tour Koenig, Pointe-aux-Sables 11134, Mauritius; dhuny@utm.ac.mu
* Correspondence: yfangli@ecust.edu.cn

Abstract: As the next frontier in wireless communication, the landscape of 6G technologies is characterized by its rapid evolution and increasing complexity, driven by the need to address global challenges such as ubiquitous connectivity, ultra-high data rates, and intelligent applications. Given the significance of 6G in shaping the future of communication and its potential to revolutionize various industries, understanding the technological evolution within this domain is crucial. Traditional topic modeling approaches fall short in adapting to the rapidly changing and highly complex nature of patent-based topic analysis in this field, thereby impeding a comprehensive understanding of the advanced technological evolution in terms of capturing temporal changes and uncovering semantic relationships. This study delves into the exploration of the evolving technologies of 6G in patent data through a novel dynamic topic modeling framework. Specifically, this work harnesses the power of large language models to effectively reduce the noise in patent data pre-processing using a prompt-based summarization technique. Then, we propose an enhanced dynamic topic modeling framework based on BERTopic to capture the time-aware features of evolving topics across periods. Additionally, we conduct comparative analysis in contextual embedding techniques and leverage SBERT pre-trained on patent data to extract the content semantics in domain-specific patent data within this framework. Finally, we apply the weak signal analysis method to identify the emerging topics in 6G technology over the periods, which makes the topic evolution analysis more interpretable than traditional topic modeling methods. The empirical results, which were validated by human experts, show that the proposed method can effectively uncover patterns of technological evolution, thus enabling its potential application to enhance strategic decision-making and stay ahead in the highly competitive and rapidly evolving technological sector.

Keywords: technological evolution analysis; dynamic topic modeling; 6G

1. Introduction

In the relentless pursuit of advancing wireless communication technologies, the transition from 5G to 6G emerges as a pivotal milestone, promising to transcend the limitations of current networks and unlock a new era of connectivity [1]. As the next frontier in wireless communication, the landscape of 6G technologies is characterized by its rapid evolution and increasing complexity, driven by the need to address global challenges such as ubiquitous connectivity, ultra-high data rates, and intelligent applications [2,3].

Unlike its predecessor, 5G, which has significantly enhanced data rates, reduced latency, and enabled massive machine-type communications, 6G is poised to deliver even more profound advancements by exploiting higher frequency bands, including the mid band (1–24 GHz), the mmWave (24–92 GHz), and sub-THz (92–300 GHz) spectrums [4]. Among these bands, the sub-terahertz band and the upper mid-band (7–24 GHz) have recently garnered significant attention as candidate frequency bands. Recent advancements in new spectrum technologies (NSTs) have highlighted the critical role of the mid-to-high frequency spectrum (7.125–24.25 GHz, FR3) as the most promising candidate for 6G deployment [5], owing to its advantages in capacity, coverage, and spectral efficiency, as well as its integration with integrated sensing and communication (ISAC) technologies [6], and it has been identified by the ITU as a critical technology for enabling key 6G applications and underscores its growing importance in the evolution of next-generation wireless systems. To enable seamless, ubiquitous connectivity in the 6G era, the integrated space–air–ground–sea communication (ISAGSC) framework has emerged as a highly promising heterogeneous network solution to address coverage challenges in harsh environments such as forests, deserts, and oceanic regions [7]. Within this framework, AI/ML-enhanced air interface (EAI) technologies have become a key research focus, with 3GPP Release 18 actively exploring the potential of AI/ML-based algorithms to enhance the air interface [8]. Current research focuses on advancements in channel state information feedback for massive multi-input multi-output, reference signals for channel estimation, beam management, and wireless positioning. In the context of communication transmission environments, reconfigurable intelligent surfaces (RISs) [9] are regarded as a transformative technology for 6G. By altering the electrical and magnetic properties of material surfaces to control the reflection of electromagnetic waves, RISs can operate without high-energy-consuming sources such as radio frequency chains, offering cost advantages over massive multi-input multi-output and providing unique wireless communication capabilities for 6G [10]. These improvements are not just incremental; they represent a paradigm shift in the capabilities and applications of wireless networks. The significance of 6G lies in its potential to catalyze the integration of communication, computing, and sensing, thereby enabling a wide array of transformative applications. Understanding the evolution of 6G technologies is crucial for researchers, industry stakeholders, and policymakers as it will shape the future landscape of communication and drive the development of new applications and services that can significantly impact various sectors of society [11]. However, the rapid proliferation and diversification of 6G technologies, coupled with their intricate interdependencies, pose significant challenges for field-related experts in manually analyzing and tracking technological evolution trends, as the dynamic landscape often outpaces traditional analytical capabilities. This highlights the necessity for more advanced and automated approaches to effectively navigate and understand the complex and fast-evolving 6G technological domain.

Analyzing patents related to 6G is an effective way to understand the complex technological landscape and fast-evolving trends in this field [11]. Patents are technical documents that contain detailed technological information and play a vital role in revealing the evolutionary trajectory of core technologies [12]. In the fast-evolving field of 6G technologies, the rapidly growing number of patent documents poses an enormous challenge for retrieving and analyzing timely information from patent data to identify technical themes and unveil technological evolution processes. In recent years, based on deep learning (DL) methods for natural language processing, many novel approaches have been developed in the field of patent analysis to reduce costs by automating tasks that previously only domain experts could solve [13]. Among them, topic modeling techniques are widely used in exploratory data analysis for organizing, understanding, and summarizing large amounts of text data [14]. Dynamic Topic Models (DTMs) are a variant of topic models designed specifically for temporal archives [15]. Unlike static topic models, such as Latent Dirichlet Allocation (LDA) [16], they possess the unique ability to adapt and refine their assessments of underlying topics as new documents are added to the corpus, allowing the identification of evolving trends and patterns in the archive [17]. This adaptability feature facilitates the identification of evolving trends and patterns in patent-based technological development [18], particularly in rapidly evolving fields like 6G. DTM-based topic evolution analysis can help track how technical concepts and innovation trends in patents change over time, thereby contributing to a more comprehensive understanding of the technological evolution in this field.

However, these approaches come with certain limitations and challenges in analyzing the evolving technology in 6G. Firstly, traditional statistics-based methods based on the Bag-of-words (BOW) assumption, such as the Term Frequency–Inverse Document Frequency model (TF-IDF) and LDA, face difficulties in effectively capturing topic semantics and dealing with high-dimensional data, resulting in potentially inaccurate and inefficient topic extraction and connection identification via similarity measurements [19]. For instance, in the early research on DTMs [15], the authors compared DTMs with static models in terms of capturing the topic evolution in Science articles. The quantitative results showed that the dynamic topic model outperformed LDA-based static topic models in predicting the content of the next year's Science articles, consistently assigning higher likelihood values. Additionally, in a study comparing LDA and BERTopic on a corpus of 22,101 climate protection articles from Austrian newspapers, LDA-like models often produced topics with very broad terminologies. For example, a housing-related topic in LDA might include general terms like 'city', 'new', and 'project', among others. At the same time, BERTopic identified more specific salient words such as 'heat pump' and 'oil heating' for the same kind of topic, making it more interpretable and better at highlighting specific aspects [20]. Nevertheless, DL-based DTM methods like BERTopic [21] handle complex semantic topic representation by utilizing high-dimensional transformer-based embedding. The study comparing LDA and BERTopic on Twitter data related to travel and COVID-19 found that LDA identified fewer topics, some of which were rather general and uninformative [17]. In dynamic topic modeling, its assumption of independent topics and reliance on word frequency might lead to overlooking topic correlations over time. For instance, it may not effectively capture how the 'government response' topic evolves in relation to other topics like 'travel restrictions' as the situation changes, making it less suitable for tracking dynamic topic changes. On the other hand, BERTopic, which uses a combination of Bidirectional Encoder Representations from Transformers (BERT) and c-TF-IDF, generated more specific and interpretable topics. For example, its c-TF-IDF algorithm can highlight the most representative terms within a cluster. In dynamic topic modeling, this feature can help in tracking how topics change over time. Despite its ability to handle complex semantic representations, significant issues arise when applying BERTopic for topic modeling in patent-related technologies, especially within the rapidly changing and complex 6G landscape. Firstly, the presence of substantial noise in the patent data, compounded by challenges in accurately understanding domain-specific technological terms, poses a significant hurdle. These factors either lead to the model's failure to cluster key technological points effectively or result in overly general clustering outcomes. Consequently, the model struggles to precisely identify and represent the core technological features within the 6G domain, undermining the effectiveness of topic-based analysis in this highly specialized and dynamic field. Additionally, detecting early signs of emerging trends in the fast-evolving landscape is crucial for strategic planning. These early signs, which are referred to as weak signals [22], indicate potential but unconfirmed changes. They may later become significant indicators of critical forces, characterized by low terms and document frequencies but high growth rates, and can signal the emergence or disruption of a target technology in 6G, even with minimal current impact. The existing DTM-based applications, such as RollingLDA [23], also face challenges in identifying and tracking emerging signals at a given moment rather than in a long-term evolution. This challenge necessitates that the model adapts to the crucial temporal dynamics of emerging trends in a complex technological landscape and demonstrates interpretability regarding shifts in industry focus. Moreover, the topics generated by these DTM-based methods often fail to provide meaningful insight for interpreting emerging technologies to enhance strategic planning.

To overcome these challenges, we propose a novel dynamic topic modeling framework in patent analysis, which employs Large Language Models (LLMs) to enhance the dynamic topic modeling framework to capture the semantics of evolving topics in a complex technological landscape and identify the emerging technology trend in a more interpretable manner. In this study, we collected 78,625 patents in the field of 6G. We performed a prompt-based technological summarization with LLMs to effectively filter out irrelevant information, thereby significantly enhancing the performance of subsequent text clustering models. Then, we conducted a comparative analysis of the performance of various embedding techniques, such as word2vec, doc2vec, BERT, and PatentsBERTa, against the baseline results and utilized contextual embedding techniques to capture semantics and extract topics in the text data. Furthermore, we adopted an enhanced BERTopic-based dynamic topic modeling framework for technological evolution analysis with random-search-based hyperparameter tuning, which can better capture the underlying temporal relationships between topics over time. Finally, to quantitatively identify the emerging trend and interpret the results to obtain technological insights in this field, we performed the weak sign analysis on the generated topics, enabling us to identify the development trends of technologies in Topic Emerging Maps (TEMs).

The main contributions of this paper are as follows:

- We proposed a novel dynamic topic modeling framework in patent analysis, which employs an enhanced BERTopic-based approach with hyperparameter optimization to capture the temporal relationship and semantics of evolving topics. Additionally, we utilized LLMs to pre-process the redundant patent data to significantly reduce the noise, thereby enabling automatic tracking of the evolving trends in a complex landscape;
- By conducting a comparative analysis of various embedding techniques, we selected the SBERT model specially pre-trained on patent data combined with temporal aligned clustering; we achieved better semantic capture and more accurate evolving topic extraction over time. This improvement enables a more in-depth understanding of the technical content within patent documents, uncovering the underlying relationships between topics over time;
- By performing a weak sign analysis on the generated topics, we demonstrate that our framework can adapt to generate more interpretable results and track technological trends in a complex landscape. The empirical studies show that our framework is an alternative tool to provide meaningful insights for industry strategic planning.

2. Literature Review

To establish the theoretical foundation for our work, we reviewed three key areas that inform our approach to analyzing 6G technological evolution through patent data.

2.1. Embedding-Based Topic Extraction for Patent Analysis

As a paradigm shift in intelligent mobile communication, 6G addresses more complex and demanding challenges, such as achieving truly global coverage, supporting ultra-high data rates, and enabling ultra-low latency communications [1,11]. Given their increasing significance and the rapid pace of technological innovation in this area, understanding the evolving technology landscape in this sector becomes essential. Patents serve as crucial references for analyzing the core technologies in this field. However, the patent text is characterized by specific, technical, legal, and, therefore, ambiguous jargon [24]. Moreover, patent texts contain novel technologies. Therefore, the traditional keyword-based approaches [25] using co-occurrence approaches, such as pointwise mutual information (PMI) or n-grams [26], often fail to adequately consider the context when defining the semantics of words and suffer from the computation expense because of the large and extremely sparse embedding matrices.

In recent years, neural network-based embedding approaches have emerged to map words or sentences into fixed-length numeric vectors. Thereby, latent semantics in massive textual data can be translated into low-dimensional dense space [27]. Word2vec and doc2vec are representative embedding techniques using neural networks based on static embedding that does not change with context once learned [28,29]. Top2vec [30], a variant of Word2Vec, capitalizes on the joint embedding of document and word semantics to find more representative and informative topics than topic models such as TF-IDF or LDA. More recently, the BERT model has utilized the advantages of LLMs and emerged as a focal point for researchers engaged in topic extraction [31]. This approach relies on dynamic embedding techniques, learning novel word vectors from the corpus by considering the contextual information. In the field of patent text mining, PatentBERT [32] and PatentSBERTa [33] focus primarily on patent text analysis and have already been shown to be very effective in these tasks. Compared to static embedding, dynamic embedding might better capture the meaning of words in a context environment; however, it may require more training time and cost.

Even though embedding-based approaches integrated with clustering techniques have been employed for topic extraction and analysis [21], there is still no consensus on evaluating these embedding methods for identifying the evolutionary states of topics in patent text analysis. Additionally, applying embedding techniques to the noise patent data without proper pre-processing may fail clustering results in the topic modeling process. In this work, we performed data pre-processing on redundant patent data using LLMs and conducted a comprehensive comparative analysis to show the performance of various embedding-based approaches to reveal the effectiveness of dynamic topic modeling for exploring technological evolution tasks.

2.2. Dynamic Topic Modeling for Analyzing Technological Evolution

Topic modeling has emerged as a powerful technique in the realm of patent analysis, offering valuable insights into the complex landscape of technological innovation. Many existing methods harness the information in patent documents to identify and classify technical topics. The typical output of a topic modeling algorithm is a set of topics, either fixed or flexible in number, with each topic represented by a list of top words [14]. DTMs introduce a novel approach by creating topics defined by a set of words and associated with a timestamp or time range [15]. They employ a Bayesian framework to model the temporal changes in the relative proportions of topics within documents. This Bayesian modeling enables the identification of evolving trends and patterns over time in the corpus [34]. Furthermore, building upon DTMs, topic evolution analysis [18] has been used in a variety of applications, such as discovering the evolution of research topics and innovations in scientific archives [35] and understanding the evolution trend in patent analysis [36].

When it comes to statistical modeling for deriving insights from data, the research of DTMs mainly falls into two categories: Probabilistic Dynamic Topic Models (PDTMs) and Algorithmic Dynamic Topic Models (ADTMs) [34]. Probabilistic topic models assume that each document in a corpus is a mixture of topics, with each topic represented as a probability distribution over the words in the corpus. PDTMs, a subset of these models, assign probabilities to different words and topics across time. This temporal probability assignment allows them to infer the underlying themes and patterns as they change [37,38]. For example, the classic DTM [15] was among the first PDTMs, which is a variant of the LDA and incorporates temporal components. Despite their extensive application across a range of studies [39,40], PDTMs exhibit elevated computational costs when confronted with substantial archives containing vast vocabularies [41]. Consequently, the work extends the class of tractable priors from Wiener processes to the more general class of Gaussian processes. This approach allows the model to be applied to large collections of text and explore topics that evolve smoothly over a long period. Despite all the efforts to address the problems of PDTMs, they still show less coherence and diversity in topic representation compared to ADTMs.

In contrast, ADTMs leverage advancements in neural networks to transform the underlying temporal probability distributions of words and documents into a fixed-length vector representation space. These models not only retain crucial words in topic descriptions but also form dense clusters of documents that are highly interpretable. For instance, BERTopic [14] is a widely-used ADTM. It leverages LLMs to generate document representations, thereby embedding context within the generated topics. This model integrates BERT and c-TF-IDF, creating dense clusters of interpretable topics while safeguarding important words in the topic descriptions. By calculating the c-TF-IDF topic representation for each time window, BERTopic enables dynamic topic modeling. This approach allows for a more nuanced understanding of how topics evolve over time, as the c-TF-IDF representation captures the relative importance of terms within each time-specific cluster, and the use of LLMs and BERT ensures that the topics are rich in context and semantic information. ANTM [34] can capture the dynamic topic over the periods using aligned UMAP to enable a flexible number of topics. Additionally, BERTrend [42] extends from BERTopic by operating within an online learning framework to dynamically classify topics.

Despite their advantages, current dynamic topic modeling approaches cannot be directly applied to identify emerging technological trends in patent analysis. Firstly, the landscape of 6G is complex and full of domain-specific technological terms. These terms often have highly specialized semantics that are difficult for existing models to interpret accurately, leading to the misclassification or omission of crucial information during topic extraction. Secondly, patent data are not only voluminous but also highly redundant. Existing dynamic topic modeling methods struggle to effectively filter out this redundancy, which can be challenging for clustering the evolution of emerging topics in the field. Moreover, the evolving technological landscape in the patent domain is characterized by rapid and discontinuous changes. With their relatively static architectures and algorithms, current models fall short of capturing these dynamic shifts in a timely and accurate manner. In contrast, we propose a novel dynamic topic modeling framework in this study, which harnesses the capabilities of LLMs to mitigate noise in redundant patent data laden with domain-specific technological terms, can better adapt to these challenges, and enables the automatic tracking of evolving technological trends within a complex technological landscape.

2.3. Weak Signal Analysis for Emerging Trend Identification

Emerging technologies in the complex landscape of 6G can have major economic impacts and affect strategic stability. The early detection of new technology trends is critical for strategic planning, as it enables users to identify opportunities and risks quickly and react to them accordingly by formulating appropriate research, development, and innovation strategies [35]. Yet, early identification of emerging technologies in large amounts of patent data remains challenging.

Early research attempted to use co-word analysis to identify emerging research themes in information security by analyzing patterns and trends [43]. Xu et al. [44] assessed the degree of topic association by considering the strength of topic linkages in multiple networks such as co-word and co-author networks. Miao et al. [35] produced a semantic analysis method to extract terms involving products, functions, and technologies from patents. They constructed the technology road mapping according to the term structure and opinions of domain experts. Park et al. [45] presented a quantitative approach to discover potential future technological opportunities from the patent–citation network. Additionally, detecting weak signals, trends, and issues in an evolving technology landscape as early as possible has been proven to be an alternative solution for strategic planning [46]. In emerging topic research, weak signals can be regarded as the early stage of emerging topics, with earliness being their primary characteristic [47]. Weak signals are identified as keywords with low frequency but high growth potential. Analyzing weak signals of emerging topics is an early identification task for specific innovative content, which can help to identify emerging topics early and provide valuable insights into their subsequent development and evolution.

Among past works about weak signal detection, many are keyword-based. Thus, portfolio maps, pioneered by the authors of [48], involve constructing Keyword Emergence Maps (KEMs) and Keyword Issue Maps (KIMs) based on two key metrics: Degree of Visibility (DoV) that quantifies the frequency of a keyword within a document set; and Degree of Diffusion (DoD) that measures the document frequency of each keyword. However, KEMs and KIMs present two major drawbacks: by focusing on keywords only, they can miss the context surrounding a weak signal, and the output is a single snapshot, which does not give clear clues of evolution over time. Maitre et al. [49] integrate LDA and Word2Vec to detect weak signals in weakly structured data. Furthermore, El Akrouchi et al. [50] introduce two functions for deep filtering metrics and the evolution of topics using numerous metrics to identify potential weak signals.

However, traditional topic modeling methods have been useful for weak signal detection but have significant limitations. They rely on pre-set topic numbers, which makes them inflexible and require hard-to-obtain prior knowledge. Also, they cannot use the context-aware embeddings from modern pre-trained models, leading to less detailed analysis. Furthermore, these methods work statically, overlooking how weak signals change over time. In identifying emerging technological trends in 6G, weak signals, as the early signs of core technologies, may emerge or fade quickly over time. Ignoring these changes can cause important trends to be misinterpreted or missed. Therefore, inspired by WISDOM [51], we propose an enhanced DTM technique with a continuous weak signal analysis method to handle emerging trends in the fast-evolving technological landscape of 6G.

3. Methodology

As illustrated in Figure 1, the methodology in this study is structured around a comprehensive DTM framework for managing evolving topics over time and exploring emerging technological topics. This framework consists of three main modules: a data acquisition and pre-processing module, a BERTopic-based DTM module, and a weak signal analysis module. Initially, the data acquisition module collected patent records from the incoPat [52] database, extracted titles and abstracts, and performed the data pre-processing process with an LLM-based summarization technique to obtain more relevant information as input without removing the technical content. Then, we deployed specifically selected pre-trained transformer-based model to generate document-level embeddings after a comparative performance analysis. Consequently, we performed a hyperparameter search to enhance the performance of the BERTopic-based framework. Finally, a weak signal analysis process was applied to interpret the weak signals in the evolving technological field and identify the emerging technological trends over time. Our automatic framework can provide valuable insights with interpretable emerging technological trends analysis for strategical decision-making in the 6G industry. The following subsections elaborate on each module's specific process and contributions in detail.

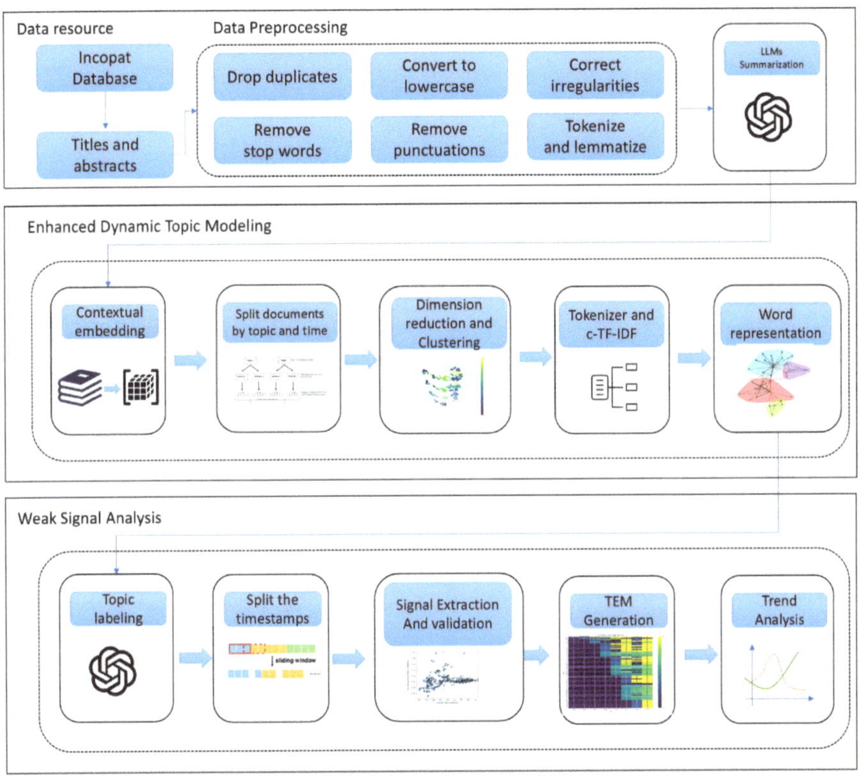

Figure 1. Our dynamic topic modeling framework for exploring the emerging technologies in 6G over time.

3.1. Data Acquisition and Text Embedding

The first component of our framework focuses on acquiring relevant patent data and transforming it into meaningful vector representations that capture semantic information.

3.1.1. Data Acquisition

We initiated our research by collecting 6G patent data in five emerging fields, namely, NSTs, RISs, ISAC, ISAGSC, and EAIs. All the related patent records were retrieved and collected from the Incopat database [52], a globally renowned patent information platform with abundant patent resources. In order to ensure a comprehensive collection of relevant patents in these fields, this study retrieved the data using specific search formulas set by experts and obtained 78,625 patent records in 6G after an initial screening process for eliminating irrelevant patents. Figure 2 shows these technological fields in 6G have been highly active in recent years and contain many emerging technologies in these five fields, which may potentially attract more industry attention. The patents' titles and abstracts, being the primary repositories of condensed yet valuable technological information, were extracted for further pre-processing.

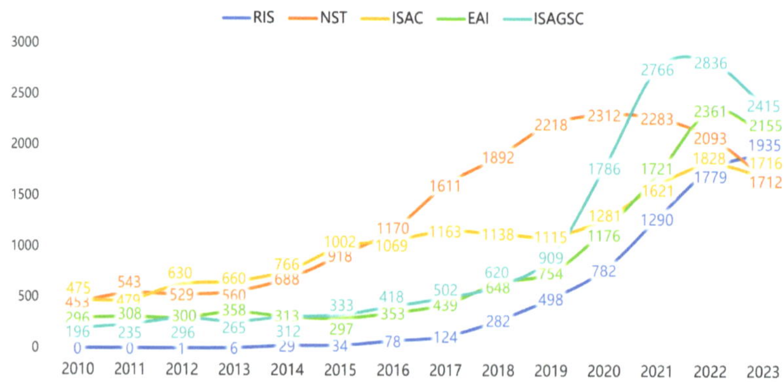

Figure 2. The detailed statistics for the five identified emerging technological fields in 6G.

Consequently, the abstracts and titles were pre-processed to reduce noise for subsequent analysis. Inspired by the work of Khandelwal [53], we employed an LLM-based summarization process to eliminate noise in the technical context and enhance the performance of dynamic topic modeling for 6G patent text analysis. In the context of dynamic topic modeling in analyzing technological trends in 6G, the patent data are often voluminous and contain a wealth of information, including technical details, background knowledge, and future research directions. Neural-based topic models like BERTopic show promise in topic modeling but can be further optimized with more concise and directed input. Secondly, the large amount of noise and domain-specific jargon in 6G patent texts can obscure the underlying themes. Summarization can help distill the essential information, reduce noise, and improve the quality of topic modeling.

3.1.2. LLM-Based Data Pre-Processing

For efficient summarization of large volumes of patent data, we employed the online API-based LLM approach ChatGLM because it allowed us to parallel process the patent context in a seamless manner. Additionally, to efficiently pre-process the large volume of patent data, we adopted an online API-based LLM approach using ChatGLM (GLM-4-Flash). This choice is because ChatGLM enables the seamless parallel processing of massive inquiries, which is more cost-effective and efficient than deploying local LLMs.

In this work, we adopted Khandelwal's [53] long and short summarization strategies using LLMs to set the different numbers of words <number> for content summarization. We further employed role-playing and few-shot in-context learning methods combined with constrained output and enhanced the prompts to further improve the performance of the DTM during the subsequent processing steps, as shown in Figure 3.

The pre-processing steps include removing hyperlinks, punctuations, and numeric values, converting text to lowercase, and eliminating unnecessary whitespaces. After that, tokenization is performed to assign unique identifiers to each unique word in the corpus.

Then, Part-Of-Speech (POS) tagging is performed to mark each token with a POS tag, like noun, adjective, verb, or adverb, according to its context, followed by lemmatization to transform tokens into their root forms. Finally, n-grams, which are continuous sequences of n words in the text corpus, are identified to uncover the language's contextual meanings.

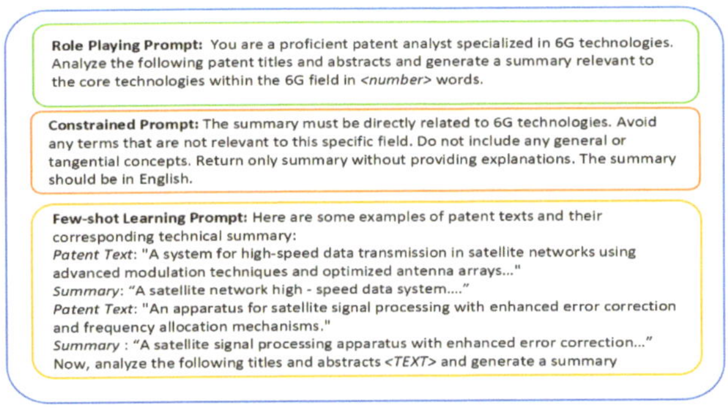

Figure 3. This figure is an example of our LLM-based summarization process for enhancing the input data in the pre-processing module.

3.2. Enhancing BERTopic-Based Framework for Dynamic Topic Modeling

Building upon the embedded representations, we developed an enhanced version of BERTopic to effectively model the temporal evolution of technological topics within the 6G domain.

3.2.1. Embedding Method Benchmark for Topic Extraction

Text embedding represents each unique word in a corpus as an N-dimensional vector. In essence, it maps words into an N-dimensional vector space, from which valuable insights, such as the semantic similarity between different words, can be deduced. When aiming to identify an effective and efficient text embedding method for unraveling dynamic technological topic evolution in 6G communication, we evaluated the performance of the aforementioned embedding approaches in terms of their ability to capture content semantics and extract technological topics. To guarantee a fair comparison during the model selection phase, we executed the document clustering tasks using the same clustering algorithm. Term Frequency–Inverse Document Frequency (TF–IDF) [54] is an initial textual embedding technique. Essentially, it is one-dimensional text embedding, which can effectively represent the importance of words in a text within the context of a document collection. However, TF-IDF has several limitations when capturing the dynamic nature of technological topics in 6G communication. Firstly, it is a static approach that only considers the frequency of words in documents without considering the semantic context in which they are used. In the rapidly evolving field of 6G, words may have different meanings depending on the specific technological context, and TF-IDF fails to capture these nuances. Furthermore, TF-IDF does not account for the emergence of new words and concepts that are characteristic of dynamic technological evolution. As 6G technology develops, new terminologies are constantly being introduced to describe novel features and technologies. TF-IDF, relying on historical document frequencies, may not be able to assign appropriate weights to these new words, potentially missing important aspects of the evolving technological topics.

In this study, we also investigated static embedding techniques, such as Doc2Vec (Word2Vec), which are based on predicting word probabilities to map words into low-dimensional dense vectors. This can be obtained through two neural network methods:

Skip Gram and CBOW. In Skip Gram, the target word is input to predict neighboring words; in CBOW, neighboring words are input to predict the target word. This way, Word2Vec can capture semantic relationships between words, placing related words closer in its vector space. However, Word2Vec has difficulty fully grasping the context-sensitive and specialized meanings of some 6G-specific terms. In addition, its embedding quality is highly dependent on the training data. In case the fast-evolving field does not cover all aspects of 6G technology, the embedding may not accurately represent word relationships.

Therefore, we further introduced contextual embedding techniques. They leverage pre-trained transformer-based LLMs, such as Data2Vec [55] and SBERT [56], to effectively capture the semantic and contextual information of words in a document. The attention mechanism in transformers allows for a more nuanced understanding of the relationships between words, enabling it to handle the complex semantic landscape of 6G technology. Furthermore, we considered the PatentSBERTa [33] in this study, which is specially trained on patent data and has never been studied in a DTM before. We conducted comparative experiments to evaluate these contextual embedding techniques and obtain a suitable embedding model for our DTM framework.

In the context of a document corpus D, each document (title and abstract) d is assigned a vector representation y. More precisely, the document embedding y for a given document d is obtained through a mapping function F, denoted as $F : d \mapsto y \in \mathbb{R}^z$. This mapping effectively captures both the contextual and semantic information from the corpus. Document embeddings are designed to represent words and documents within a low-dimensional feature vector space. In this space, the embedding dimension z is significantly smaller than the vocabulary size (i.e., the count of distinct words in D). This dimensionality reduction is a key characteristic of word embedding methodologies as it enables more efficient processing and representation of text data while still retaining essential semantic and syntactic information. It allows for better generalization and comparison of documents, facilitating topic modeling tasks for 6G.

3.2.2. Hyperparameter Tuning for BERTopic

In the rapidly evolving landscape of 6G technology, the continuous tracking of the 'topic evolution' is crucial for the understanding of emerging technologies in 6G. Traditional topic modeling approaches are naturally static and thus do not allow for the modeling of documents ordered sequentially. It is understood that dealing with dynamic, constantly changing volumes of patent data would require transitioning to a DTM. Given the cyclic nature of topic evolution, we adopt the BERTopic-based framework, as shown in the second part of Figure 1.

Initially, BERTopic was fitted as if the data have no temporal aspect, thereby creating a general topic model. After obtaining the embeddings from contextual embeddings, the three critical components used in our BERTopic-based framework are as follows:

1. We first performed the Uniform Manifold Approximation and Projection (UMAP) algorithm to reduce their dimensionality. The high-dimensional vectors obtained from the embedding model are often redundant and computationally expensive to process, especially in the subsequent clustering step. UMAP aims to find a low-dimensional representation of the data that preserves the local and global structure of high-dimensional data. It constructs a fuzzy topological representation of the data in the high-dimensional space and then optimizes a low-dimensional representation approximating this topology. In processing 6G patent data, UMAP can transform the high-dimensional vectors (e.g., vectors in $\mathbb{R}^z$) into two- or three-dimensional vectors (e.g., in $\mathbb{R}^2$ or $\mathbb{R}^3$), which can significantly reduce the computational complexity of the subsequent clustering step and is easy to visualize;
2. Hierarchical Density-Based Spatial Clustering of Applications with Noise (HDBSCAN) was used to group the low-dimensional vectors obtained from UMAP into clusters.

HDBSCAN is a hierarchical and density-based clustering method that allows it to handle clusters of irregular shapes. The topics in the 6G landscape may have complex and non-spherical distributions in the low-dimensional space, and HDBSCAN can effectively identify these clusters. Let V' be the set of low-dimensional vectors obtained from UMAP and HDBSCAN partition V' into k clusters $C_1, \ldots, C_k$ and a set of outliers O;

3. Cluster tagging using c-TF-IDF is the final step in BERTopic to extract topic representation for each cluster at different time frames. Global representation was utilized to identify the main topics likely to emerge at different timesteps. For each topic and timestep, the c-TF-IDF representation was calculated to identify representative terms for each cluster. The mathematical details of c-TF-IDF are presented in the upcoming section. In the tasks of the DTM, both global fine-tuning and evolutionary fine-tuning were used to enhance the results. Firstly, global fine-tuning was carried out. This method combines the c-TF-IDF representation of a topic at a particular timestep with the global representation. By taking an average of these two representations, the topic representation at that timestep was adjusted to move slightly closer to the global one. This approach enables the topic to maintain some of its unique characteristics while also aligning with the overall trends captured by the global representation. Secondly, evolutionary fine-tuning was performed. Here, the c-TF-IDF representation of a topic at a given timestep was averaged with the c-TF-IDF representation of the same topic at the previous timestep. This process allows the topic representation to gradually change and adapt over time, reflecting the dynamic nature of the data. As new information becomes available at each timestep, the topic representation evolves to incorporate these changes. These keywords effectively summarize the main concepts and characteristics of the cluster, providing valuable insights into the technological topics within that time period;
4. Finally, integrating KeyBERT for keyword extraction with the Maximal Representation Model yielded promising results. KeyBERT, leveraging pre-trained language models like BERT, can effectively identify the most salient keywords within the text. In the context of 6G patent emerging technologies analysis, these keywords are crucial as they distill the essence of the patent content. When combined with the Maximal Representation Model, the extracted keywords enhance the model's ability to prioritize informative content. The Maximal Representation Model aims to capture the most significant aspects of the data. By feeding the keywords identified by KeyBERT into this model, it can better focus on the important terms within the data.

Additionally, we performed a random search for hyperparameter optimization to select the hyperparameters of the UMAP and HDBSCAN models based on the DBCV value [57], which is a score for the goodness of a density-based clustering model spanning $[-1, 1]$. During each iteration of the random search, we randomly sampled 25% of the abstracts from the data prepared. In the optimal run, we achieved a DBCV score of 0.45. More detailed parameter settings are explained in the Experimental Results section.

3.3. *Topic Representation and Emerging Trend Analysis*

After extracting the evolving topics, we employed specialized techniques to represent them meaningfully and analyze emerging technological trends within the 6G landscape.

3.3.1. Topic Representation

To obtain representative technical keywords for each cluster in each time frame within documents, the topic representation method was employed. This method is crucial for identifying the most relevant terms or phrases that characterize a particular concept in each time frame.

We adopted the $c-TF-IDF$, a variation of $TF-IDF$ that considers the class labels of documents, to compute a list of m representative terms describing each document cluster's semantic contents. $TF-IDF$ calculates the importance of a word w in a document d within a collection of documents D as follows:

$$c-TF-IDF(w,d,c,D) = TF(w,d) \times IDF_c(w,D) \tag{1}$$

where $TF(w,d)$ is the term frequency of word w in a document d. $IDF_C(w,D)$ is the inverse document frequency of word w in a collection of documents D in class c. It is calculated as follows:

$$IDF_c(w,D) = \log \frac{|D_c|}{|d \in D_c : w \in d|} \tag{2}$$

where $|Dc|$ is the total number of documents in class c. We used $c-TF-IDF$ to compute a list of representative terms describing each document cluster's semantic contents. For a given document cluster in a particular time frame t, we calculated the $c-TF-IDF$ values for all words in the documents within the local cluster. Then, we sorted these words in descending order based on their $c-TF-IDF$ values. The top words were selected as the representative technical key words for that cluster in the time frame. Therefore, these key words can effectively summarize the main concepts and characteristics of the cluster, providing valuable insights into the technological topics within that time period.

3.3.2. Emerging Trend Analysis

The early identification of emerging signals and trends in the constantly evolving technology landscape of 6G represents a substantial challenge. By harnessing the concept of weak signals within the extracted topics, this study is dedicated to pinpointing technological topics currently garnering limited attention yet possessing substantial promise for future development. In the realm of 6G patent research, discerning the annual evolution of key technologies is pivotal for foreseeing industry advancements. After leveraging BERTopic to extract evolving topics from 6G patent data, we adopted the Topic Emergence Map (TEM) approach to analyze the emerging trends in the topic evolution. By visualizing each topic's signal state across different years as a heatmap, we can gain more intuitive insights into the development of 6G technology topics and detect emerging trends.

In order to quantitatively extract future signals and make the result more interpretable, an early study [48] inspired us to employ topic proportions obtained from our time-aware dynamic topic modeling framework and use TEMs to identify future signs of emerging technologies in related 6G fields.

The TEM is a valuable method for identifying future-oriented signals within research topics. Its underlying principle lies in recognizing that the early detection of emerging signals and trends is crucial in a rapidly evolving technological landscape. The TEM is constructed using two key dimensions. The *x-axis* represents the average topic proportions derived from the BERTopic model, which reflects a topic's relative significance in the dataset at a given time frame. The *y-axis*, on the other hand, denotes the rate of increase in topic proportions, measuring how quickly a topic's importance changes over time. Based on the mean of the *x-axis* values and a *y-axis* value of zero, the TEM is partitioned into four quadrants: strong signals (high average proportion and high growth rate), weak signals (low average proportion but high growth rate), latent signals (low average proportion and low growth rate), and not strong but well-known (NSWK) signals (high average proportion and low growth rate).

In order to utilize the TEM principle to create a heat map to analyze the evolution of 6G technology topics, a series of steps are involved. First, the 6G patent data, which have been processed with topics extracted by BERTopic, needs further processing. Each topic should be associated with its corresponding patent documents and timestamped according to the patent application year. Then, calculate the topic proportions for each year by dividing the number of patent documents of a specific topic in a year by the total number of patents in that year. Simultaneously, the rate of increase in topic proportions for consecutive years is determined. Afterward, a matrix with rows representing different 6G technology topics and columns representing different years is created. Color each cell in the matrix based on the topic's position in the TEM quadrant in the corresponding year. Finally, data visualization tools are employed to generate the heatmap. Observing the heatmap allows us to analyze the evolution of 6G technology topics, identify trends, and compare different topics in the same year, enabling a comprehensive understanding of the 6G technology landscape over time.

We segmented the examined period into three-year intervals with one year overlapping to track the temporal evolution and changes in signal patterns. We refer to these intervals as $\{pi\}$. TEMs were then built for each period, with a signal label assigned to each topic within the TEM across the examined periods.

4. Experimental Results and Comparative Analyses

The experiments conducted in this study were designed to comprehensively validate the three key components of the methodology proposed in Section 3. Given the dynamic and complex nature of 6G technology and the limitations of existing research methods in accurately analyzing its technological evolution, our experiments aimed to provide robust evidence for the effectiveness of our approach.

4.1. Datasets and Experimental Setup

Our experimental evaluation begins with describing the datasets used and the specific parameters and metrics employed to assess model performance.

4.1.1. Datasets

To explore the technological evolution in the domain-specific fields in 6G, we collected the datasets of five emerging fields in 6G using a carefully crafted search query developed by domain experts. These fields include new spectrum technologies (NSTs), reconfigurable intelligent surfaces (RISs), integrated sensing and communication (ISAC), integrated space–air–ground–sea communication (ISAGSC), and enhanced air interfaces (EAIs). The detailed statistics of the datasets are illustrated in Table 1. Given the restricted number of topics suitable for modeling in the early years, the time period from 2016 to 2024 was selected for performing the DTM. This time frame provides a more comprehensive and representative data source for the research, compensating for the limitations in the initial stages and enabling a more in-depth exploration of the technological evolution in these 6G fields.

Table 1. The statistics of 6G patent datasets used in the experiment.

Patent Subjects	Periods	Documents	Vocabulary	Tokens
NSTs	2016–2024	22,834	18,945	16.952 MB
RISs	2016–2024	7863	9705	6.070 MB
ISAC	2016–2024	14,918	21,715	15.055 MB
ISAGSC	2016–2024	17,240	14,918	12.462 MB
EAIs	2016–2024	15,770	17,876	11.386 MB

4.1.2. Evaluation Metric

As the number of topics produced by a dynamic topic model does not adequately gauge its quality or superiority compared to other models, we assessed the performance of the topic models from the perspectives of human interpretability and diversity. Specifically, we adopted two widely used metrics, namely, topic coherence and topic diversity, for this evaluation.

Topic coherence (TC) [58], referring to the coherence of top words in a topic, indicates the interpretability of a topic containing m words. If the words in a topic often appear together in the same documents, it indicates that the topic has a high coherence. We employed the popular the co-occurrence metric CV as the coherence metric [59], which outperforms early metrics like Normalized Pointwise Mutual Information (NPMI) [60], which is an improved version of PMI [61]. The formula for PMI is as follows:

$$PMI(w_i, w_j) = log\frac{P(w_i, w_j)}{P(w_i)P(w_j)} \tag{3}$$

where $P(w_i)$ and $P(w_j)$ are the probabilities of words w_i and w_j being present in the window, and $P(w_i, w_j)$ is the probability that both are present in the window. $NPMI$ normalizes PMI to keep the values between -1 and 1, potentiating the result for better differentiation. Its formula is:

$$NPMI(w_i, w_j)^\gamma = (\frac{PMI(w_i, w_j)}{-log(P(w_i, w_j))})^\gamma \tag{4}$$

The context vector for a word w_i is then

$$\vec{v}(w_i) = \left\{ \sum_{w_j \in W} NPMI(w_i, w_j)^\gamma \right\} \tag{5}$$

where W is the vocabulary. For a given topic, the cosine similarity between the context vectors of all its top words was calculated. After that, all these distances were aggregated, and the average value is the topic coherence Cv. To measure the association between a topic and its time slice, we also used the documents within that time slice as a reference corpus to estimate the probabilities of word occurrences. In practice, a higher TC value implies that the topic exhibits greater coherence within the documents of the time slice.

Topic diversity (TD) [62] is used to evaluate the degree of difference in representing a given set of topics. For a given topic within a specific time slice, we calculated the proportion of its 15 top-ranked words that both occur only once and are present within that same time slice. TD is defined as the percentage of unique words in the top N words of the topics. The formula for TD is relatively straightforward:

$$TD = \frac{\text{Number of unique words in top } N \text{ words}}{\text{Total number of top } N \text{ words}} \tag{6}$$

TD is used to evaluate the distinctiveness of topics. A TD value close to 0 suggests many repeated words among the topics, indicating redundant topics. On the other hand, a value close to 1 implies that the topics are more singular, with a high proportion of unique words. In a news article corpus, high TD values would mean that different topics cover a wide range of distinct concepts, while low TD values might indicate that topics are overlapping and less diverse. Consequently, TD can be used to determine whether repetitive topics exist. This metric is useful for ensuring that the topics generated by a topic model are diverse enough to represent different aspects of the corpus content.

4.2. Performance Evaluations

In this section, we present comprehensive empirical findings from our experiments, analyzing the effects of different components of our methodology on topic modeling performance across the 6G patent datasets.

4.2.1. Comparisons of LLM-Based Pre-Processing Techniques

In our research on 6G patent data, the primary objective was to understand how different summarization strategies impact the performance of various tasks related to 6G technological analysis. We collected patent datasets from five emerging fields in 6G from 2016 to 2024. To process this large volume of patent data, we employed ChatGLM, an online API-based LLM protocol. It was selected for its ability to seamlessly parallel process the patent context, making it more cost-effective and efficient than local LLMs. For content summarization, we adopted long and short summarization strategies with different constraints on the number of words, which were 15–30 and 30–60, respectively. Additionally, we used refined prompts as proposed in Section 3.1.2 to further improve the results and prevent irrelevant information in the topic extractions. We adopted the all-MiniLM-L6-v2, a typical SBERT model, as the default baseline setting in this experiment. We calculated the TC and TD metrics for each period at one-year intervals throughout the entire duration. The values reported for both TC and TD are the average values calculated across all the years.

From the data presented in Table 2 and Figure 4, it is evident that the application of our proposed summarization techniques, when combined with our LLM-based pre-processing, has distinct effects on the performance metrics of TC and TD across different datasets consistently.

Table 2. This table compares the performance of annual TC and TD averages in DTM (From 2016 to 2024). We compare the baseline to our LLM-based pre-processing techniques with long and short summarization techniques. The elements highlighted in red in the table represent the maximum values in the corresponding columns.

Model	Metrics	NSTs	RISs	ISAC	ISAGSC	EAIs
Baseline	TC	0.4900	0.4978	0.4749	0.4683	0.4615
	TD	0.5968	0.6849	0.6775	0.6542	0.6647
Baseline + long sum	TC	0.5039	0.5002	0.5024	0.4883	0.4717
	TD	0.6157	0.6985	0.6923	0.6643	0.6983
Baseline + short sum	TC	0.4005	0.4314	0.4140	0.4062	0.4032
	TD	0.6656	0.7159	0.7537	0.6923	0.7400

For the TC metric in Table 2, compared to the baseline, the 'Baseline + long-sum' approach shows an improvement in most fields in five datasets. For instance, in the NST field, the TC value increased from 0.4900 to 0.5039, indicating that the long summarization (a key element of the 'Baseline + long - sum' approach, characterized by its ability to preserve detailed information) indeed helps in more complex analysis tasks. Similarly, in the ISAC field, the value rose from 0.4749 to 0.5024. This improvement is consistent with our previous analysis that long summarization are beneficial for exploring complex relationships within 6G technological concepts. However, the 'Baseline + short-sum' approach led to a decrease in TC performance across all fields. This performance decrease is expected as short summarization (a key element of the 'Baseline + short - sum' approach) sacrifice detailed information crucial for complexity-related tasks.

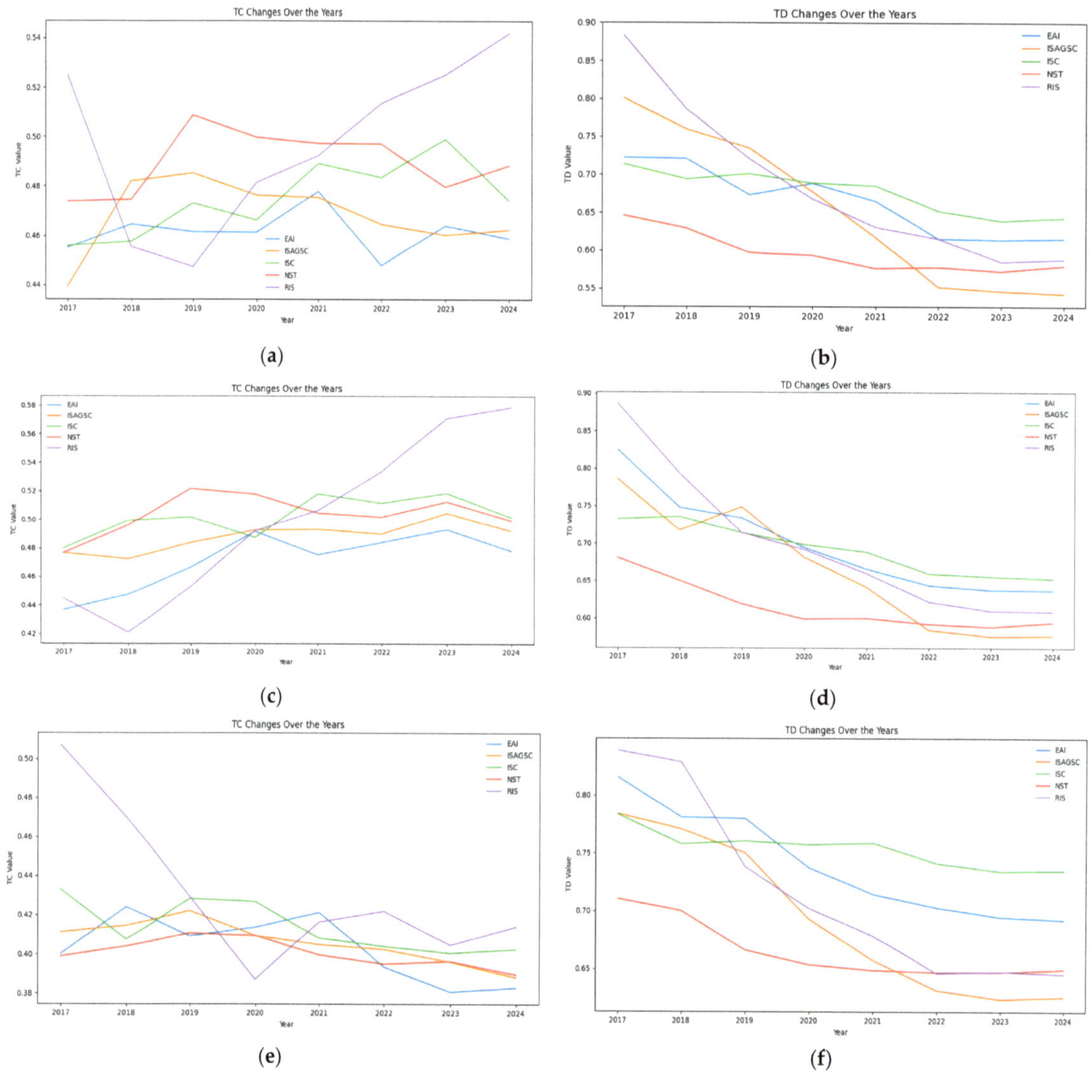

Figure 4. TC (left-hand side panels: (**a**) baseline, (**c**) baseline + long-sum, and (**e**) baseline + short-sum) values and TD (right-hand side panels: (**b**) baseline, (**d**) baseline + long-sum, and (**f**) baseline + short-sum) values over the years for different topic modeling approaches. The baseline represents the baseline model, while the long and short models are our LLM-based summarization techniques with long-and short-length summaries, respectively, illustrating the performance trends in TC and TD metrics over the years.

The baseline model's TC curve in Figure 4 shows a trend over the years, potentially reflecting the baseline performance in capturing topic-related characteristics. The long-sum and short-sum models, which are based on BERTopic with different summary lengths, exhibit distinct TC patterns. These differences might indicate that the summary length impacts how well the model can represent topics in terms of TC, with variations in the curves suggesting different levels of effectiveness or adaptability to the data over time.

Regarding the TD metric in Table 2, the 'Baseline + short-sum' approach demonstrates significant enhancements. In the ISAC field, the TD value increased from 0.6775 to 0.7537, and in the EAI field, it rose from 0.6647 to 0.7400. This clearly shows that short summarization, which focuses on key point extraction, is highly effective for density-related tasks. In contrast, the 'Baseline + long-sum' approach also improved the TD performance in some fields like EAIs, increasing from 0.6647 to 0.6983. However, the overall improvement was more pronounced with the short-sum method.

In the Supplementary Materials, we included one folder named 'Heatmaps' to comprehensively visualize the performance of different embedding methods. These visualizations are presented for each dataset under three conditions: the baseline model, the baseline with short summarization (baseline + short sum), and the baseline with long summarization (baseline + long sum). The heatmaps specifically focus on cosine similarity, which is a crucial metric for evaluating the embedding quality. By comparing the cosine similarity heatmaps across these three scenarios, we can clearly observe the differences in how well each approach captures semantic relationships within the patent data. For example, we can observe how the long summarization might preserve more detailed semantic information, leading to higher cosine similarity values in relevant regions of the heatmap compared to the baseline method. This provides an intuitive understanding of the impact of different summarization strategies and the baseline model on the quality of embeddings, further validating the effectiveness of our proposed methods in handling 6G patent data. Additionally, to further validate the effectiveness of our LLM-based pre-processing methods, we conducted additional experiments and included 2D visualizations of the embeddings for each dataset in the Supplementary Materials named 'Clusters'. We utilized UMAP, a powerful dimensionality reduction technique, to project the high-dimensional embeddings into a 2D space for better visualization. The generated HTML-based interactive visualizations offer an intuitive understanding of the impact of LLM-based pre-processing on the patent data. In these visualizations, we can clearly observe the distribution of data points. When LLM-based pre-processing was applied, the data points tended to cluster more tightly, indicating a reduction in noise. For instance, in the RIS dataset, the pre-processed data (both the long summarization approach and short summarization approach) show a more organized structure, with data points forming distinct groups. This suggests that LLM-based summarization effectively filters out irrelevant information, enabling the model to better capture the underlying semantic relationships within the patent data. These visualizations serve as evidence for the noise-reducing capabilities of our LLM-based pre-processing approach, complementing our overall analysis of the dynamic topic modeling framework for 6G patent data.

In summary, our research validates that different LLM-based summarization strategies have specific and predictable impacts on 6G-related task performances. The long-summarization technique enhances the overall performance compared to the baseline, making it a more fitting choice for complex analysis tasks that demand in-depth understanding and comprehensive interpretation. In contrast, short-summarization techniques stand out in key point extraction, where the goal is to quickly distill the most crucial information and high-speed process tasks, enabling rapid and efficient handling of large volumes of text. These findings can guide future research and application development in 6G patent data analysis, enabling more efficient and accurate utilization of the vast amount of patent information in the 6G domain.

4.2.2. Comparison of Contextual Embedding Techniques

Our experimental study focused on evaluating the performance of different contextual embedding models in a DTM in 6G. We investigated four models: the Baseline (SBERT model: all-MiniLM-L6-v2), PatentSBerta [33], SciBert [63], and Roberta [64]. The comparison was conducted across five parameters: NSTs, RISs, ISAC, ISAGSC, and EAIs. After analyzing the data presented in the two tables (Tables 3 and 4), it is clear that no single model can be declared an absolute winner across all the evaluated parameters.

Table 3. This table shows the comparison of performances of averages of TC in different models. The elements highlighted in red in the table represent the maximum values in the corresponding columns.

Model	NSTs	RISs	ISAC	ISAGSC	EAIs
Baseline	0.4900	0.4978	0.4749	0.4683	0.4615
PatentSBerta	0.4823	0.4953	0.4771	0.4768	0.4532
SciBert	0.4875	0.5090	0.4812	0.4709	0.4571
Roberta	0.4848	0.4967	0.4835	0.4675	0.4615

Table 4. This table shows the comparison of performances of averages of TD in different models. The elements highlighted in red in the table represent the maximum values in the corresponding columns.

Model	NSTs	RISs	ISAC	ISAGSC	EAIs
Baseline	0.5968	0.6849	0.6775	0.6542	0.6647
PatentSBerta	0.5860	0.6813	0.6739	0.6480	0.6645
SciBert	0.5999	0.6735	0.6633	0.6508	0.6733
Roberta	0.5922	0.6703	0.6755	0.6473	0.6724

The baseline model shows a relatively high NST score of 0.4900, yet it does not consistently outperform other metrics. In some cases, PatentSBerta has slightly lower NSTs but higher ISAC than the baseline. SciBert demonstrates strong performance in RISs with a value of 0.5090, but its performance in other metrics is not uniformly superior. Roberta also has a mixed bag of results, with some metrics showing comparable values to the others.

This lack of a winner indicates that the choice of model depends on the specific metric of interest. However, if overall balanced performance is desired, a more in-depth analysis of each metric's weights and importance would be necessary. Moreover, these results suggest that there is room for further improvement in developing these models to achieve more consistent and superior performance across all the evaluated metrics.

4.2.3. Results of Hyperparameter Tuning

We further tested our model with hyperparameter tuning, and the experimental results indicate that the optimized hyperparameters significantly enhanced the model's performance. After conducting random searches based on the DBCV value, the carefully selected parameters for the UMAP and HDBSCAN models brought about substantial improvements in the clustering quality.

For the UMAP model, setting n_neighbors = 50, n_components = 50, min_dist = 0, and metric = 'cosine' enabled a more effective dimensionality reduction. The reduced-dimensional vectors retained the essential structure of the high-dimensional data and facilitated the subsequent clustering process. This was evident in the improved performance of the HDBSCAN model, which was able to identify more distinct and meaningful clusters.

With the HDBSCAN model's parameters set as min_cluster_size = 100, min_samples = 10, metric = 'euclidean', and cluster_selection_method = 'leaf', the model could better handle the complex distribution of 6G patent data. It effectively separated the data into clusters, minimizing the number of misclassified outliers. As a result, the topic representation obtained from the c-TF-IDF step became more accurate and interpretable.

In terms of the evaluation metrics, as shown in Table 5, the optimized model showed a remarkable increase in topic coherence. The top words selected for each cluster were more closely related semantically, with the average topic coherence (TC) value increasing by approximately 40%, on average, across all datasets compared to the non-optimized model. This improvement in TC indicates that the model better captured the underlying concepts within each cluster.

Table 5. This table shows the comparison of performances of averages of TC in the baseline model and baseline with hyperparameter optimizations. The elements highlighted in red in the table represent the maximum values in the corresponding columns.

Model	NSTs	RISs	ISAC	ISAGSC	EAIs
Baseline	0.4900	0.4978	0.4749	0.4683	0.4615
Baseline + OPT	0.7877	0.6511	0.5498	0.6412	0.5982

Simultaneously, the topic diversity (TD) also witnessed a positive change, as shown in Table 6. The proportion of unique words in the top-ranked words of each topic increased, resulting in a more diverse set of topics. This was particularly beneficial for comprehensively representing the various technological aspects in the 6G patent data. The TD value increased by about 30%, on average, suggesting that the hyperparameter-tuned model could generate more distinct and non-redundant topics.

Table 6. This table shows the comparisons of performances of averages of TD in the baseline model and baseline with hyperparameter optimizations. The elements highlighted in red in the table represent the maximum values in the corresponding columns.

Model	NSTs	RISs	ISAC	ISAGSC	EAIs
Baseline	0.5968	0.6849	0.6775	0.6542	0.6647
Baseline + OPT	0.8898	0.9080	0.9467	0.9241	0.9463

In summary, the hyperparameter tuning process was crucial for achieving better performance in our 6G technology DTM model. It provided a more refined way of handling patent data in a complex landscape, leading to more accurate topic representation and a deeper understanding of the emerging trends in 6G technology. Future research could explore more advanced hyperparameter optimization techniques, such as random searches or Bayesian optimization, to further improve the model's performance and generalization ability.

4.3. Empirical Analysis

In the empirical analysis section of this study, our experiments were designed to comprehensively explore the technological evolution in the 6G field. We conducted DTM analysis on individual subject fields of five datasets in 6G, which included NSTs, RISs, ISAC, ISAGSC, and EAIs. Based on the concept of weak signal analysis, we calculated the TEM and heatmap to quantitatively identify the weak signals of 6G, which is an emerging technology in the near future. These results provide a deeper understanding of the technological evolution in the 6G field, enabling us to identify emerging technologies and trends. This information can support strategic decision-making in the industry, helping stakeholders to plan research and development directions and stay ahead in the highly competitive 6G market.

Firstly, we adopted a BERTopic-based DTM to conduct an analysis of the overall trends for 6G technology. Figure 5 shows the top-ranked topics from the charts for five datasets. The topics with line chart visualizations were generated based on our DTM framework.

Clearly, all the technologies that are displayed as top-rated topics have grown fast over the years. As we can see in the charts, in the new spectrum technology (NST) of 6G, the 'power amplifier_mixer_inductor_stage' has been rated as the top-most ranked topic recently. This is likely because the power amplifier is essential for boosting the signal power. Since 6G requires high-speed data transmission over long distances in a wider spectrum, a power amplifier can ensure that the signal can still be accurately detected after suffering from attenuation during propagation. The mixer is mainly responsible for frequency conversion. It enables the transformation of signals between different frequencies, which is beneficial for signal processing and transmission in the 6G new spectrum. Inductors, on the other hand, work with capacitors to form LC filters, filtering out unwanted signals and noise. They also play a role in energy storage and impedance matching. Similarly, we can easily identify the top-rated topics for the remaining four subjects from Figure 5, which are the 'dielectric substrate_metasurface antenna' for RISs, 'radar communication_ofdm' for ISAC, 'satellite_network' for ISAGSC, and 'beam management' for EAIs.

To further explore the technological evolution in the domain-specific fields in 6G, we employed TEMs associated with heatmaps to quantify the emergence of technological shifts in the five subjects of the patent datasets, as shown in Figure 6. The weak signals refer to the signals with a low average proportion but a high growth rate in the obtained topics, as shown in Table 7.

For instance, in the NST dataset, the identified weak signals provide valuable insights into the emerging technological directions. Topic 17, '17_channel estimation_sparse_millimeter wave channel_deep learning', indicates that deep learning-based methods for sparse millimeter-wave channel estimation are emerging. In the context of 6G, millimeter-wave communication is crucial for achieving high-data-rate transmission. However, the complex channel environment poses challenges. Deep learning can potentially optimize channel estimation algorithms, improving signal reception and transmission quality. Topic 18, '18_millimeter wave radar_wave radar_doppler_target detection', suggests that millimeter-wave radar technology for Doppler target detection is on the rise. This is important for applications such as 6G-enabled autonomous driving, where precise target detection is essential for vehicle safety. By observing the TEM heatmap for NSTs, we can see how these topics' importance has changed over time. A topic with a low average proportion but a high growth rate, like these two, is considered a weak signal. Monitoring such topics helps stakeholders in the 6G industry anticipate future technological needs and invest in relevant research and development.

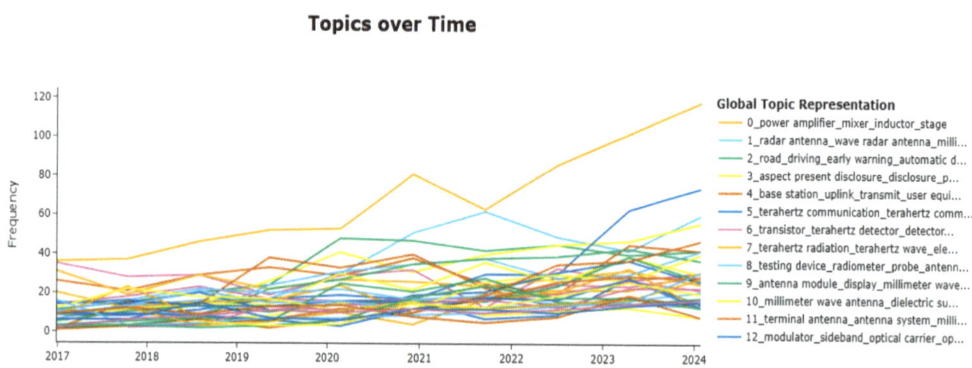

Figure 5. *Cont.*

(a) Overall trends of NSTs

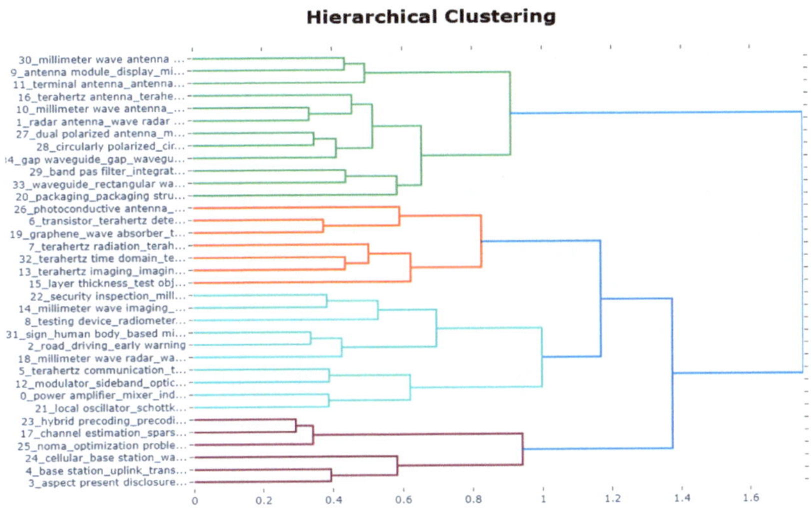

(b) Hierarchy clustering for NSTs

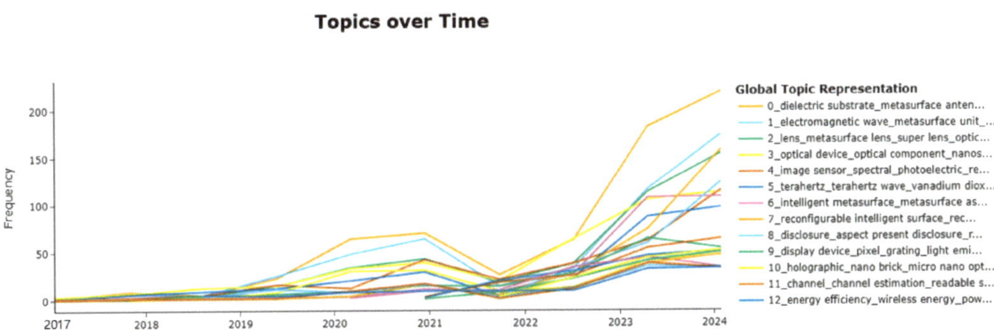

(c) Overall trends of RISs

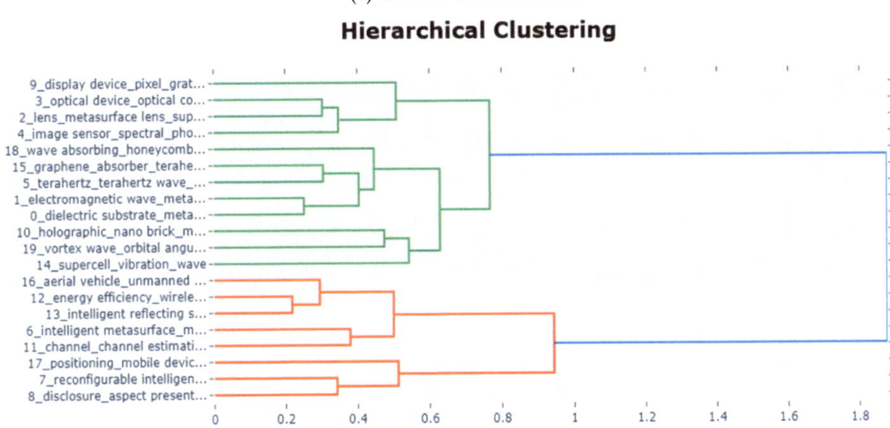

(d) Hierarchy clustering for RISs

Figure 5. *Cont.*

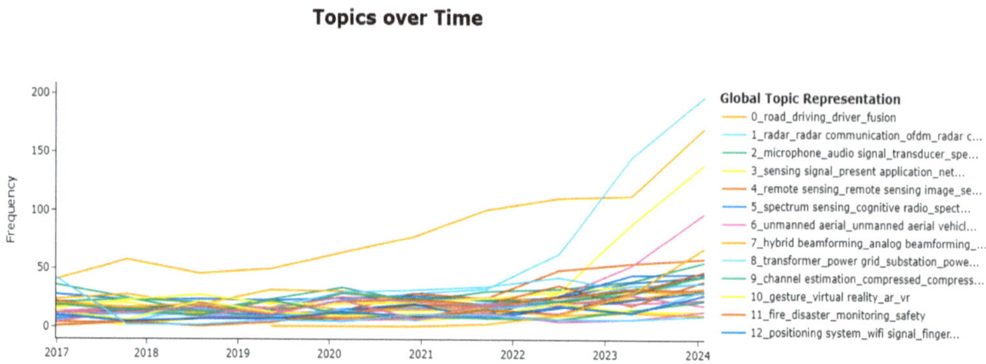

(**e**) Overall trends of ISAC

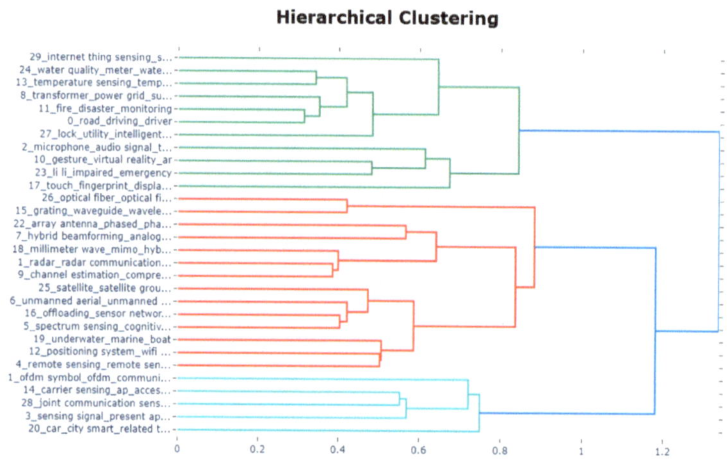

(**f**) Hierarchy clustering of ISAC

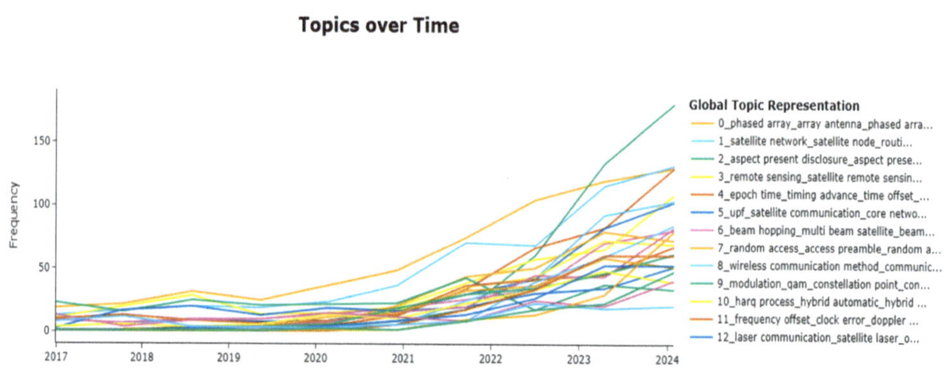

(**g**) Overall trends of ISAGSC

Figure 5. *Cont.*

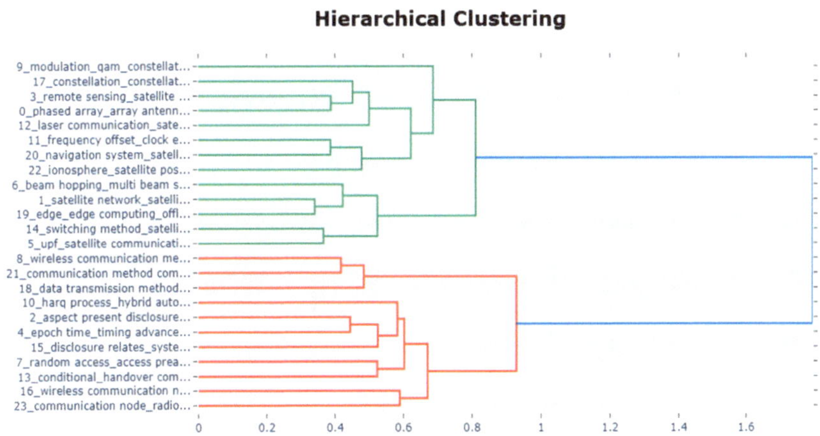

(**h**) Hierarchy clustering of ISAGSC

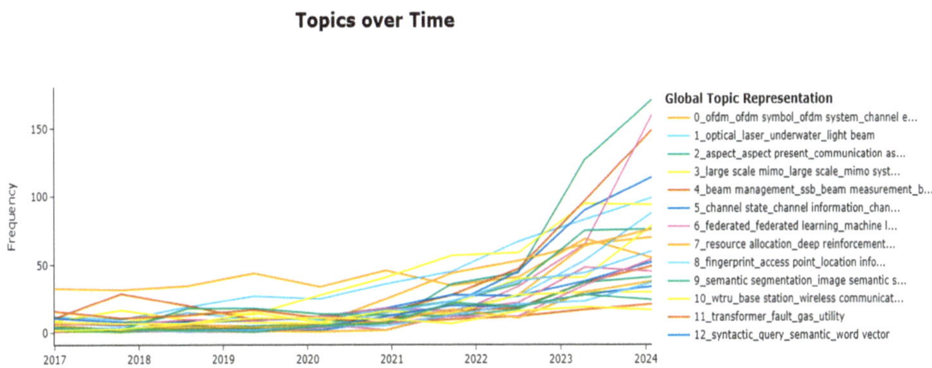

(**i**) Overall trends of EAIs

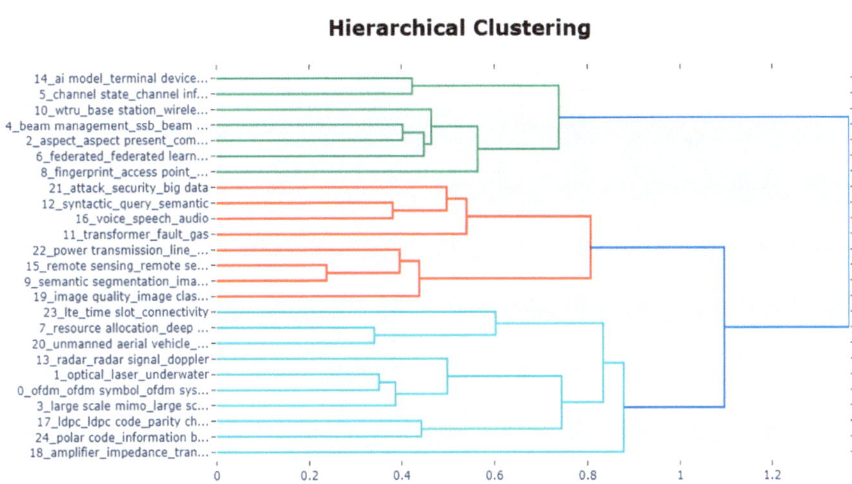

(**j**) Hierarchy clustering of EAIs

Figure 5. This figure shows overall trends and hierarchy clustering for 5 subjects.

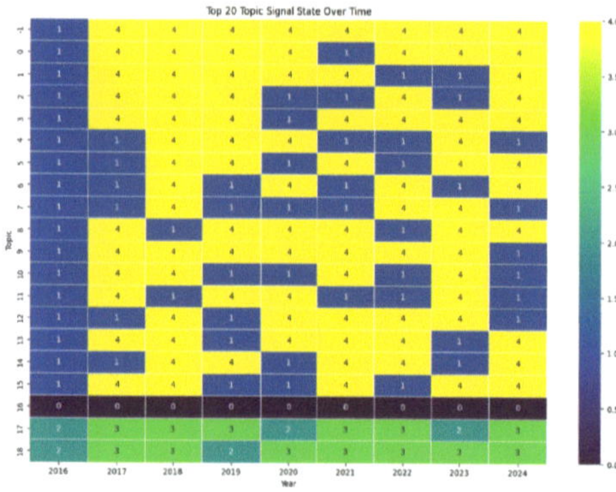

(**a**) Evolution heatmap of NSTs

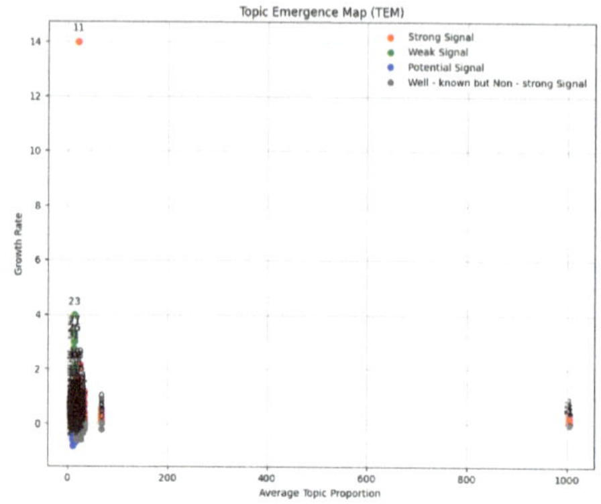

(**b**) TEM for NSTs

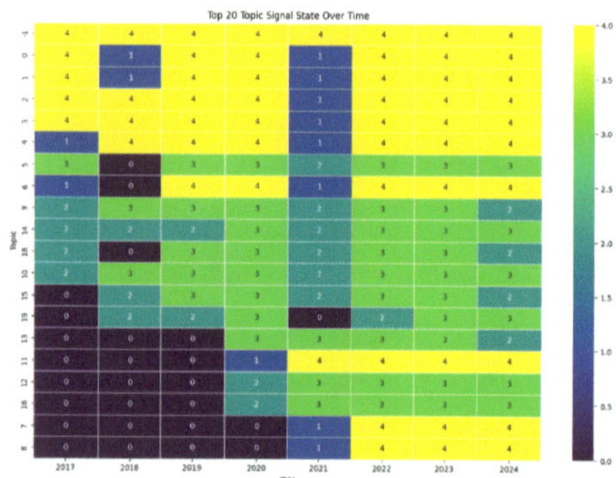

(**c**) Evolution heatmap of RISs

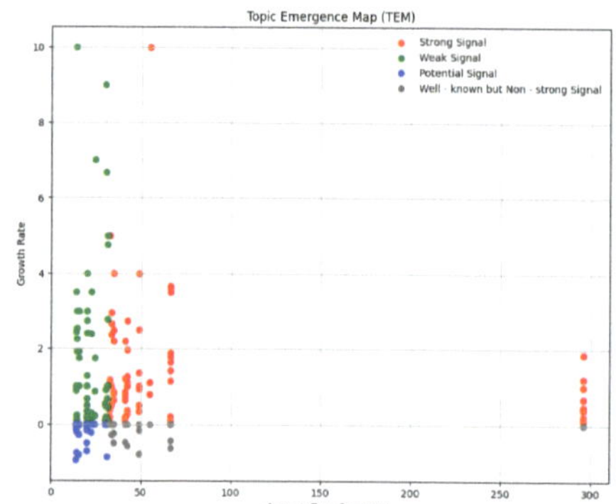

(**d**) TEM for RISs

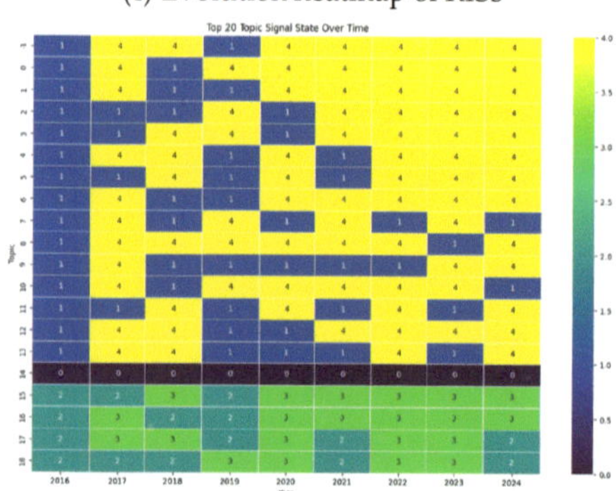

(**e**) Overall trends of ISAC

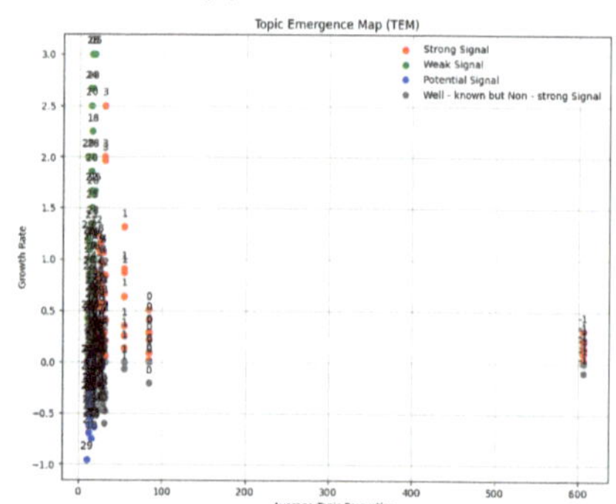

(**f**) Hierarchy clustering of ISAC

Figure 6. *Cont.*

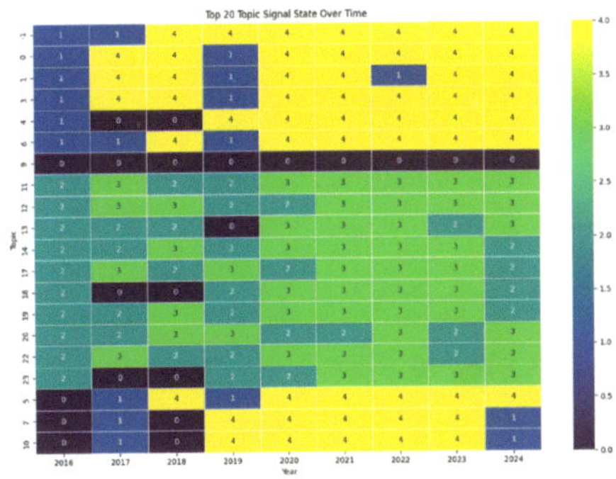

(**g**) Overall trends of ISAGSC

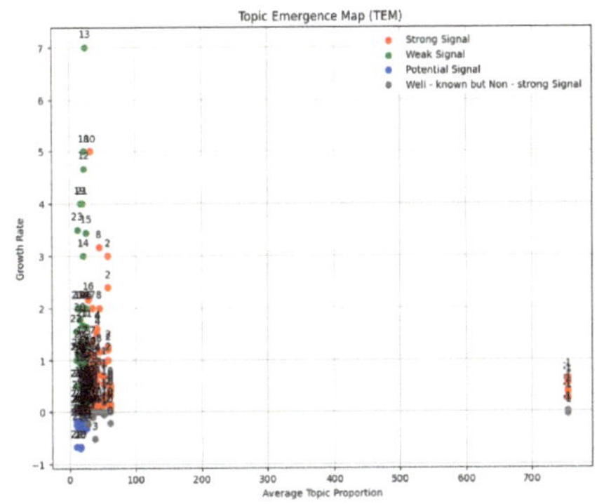

(**h**) Hierarchy clustering of ISAGSC

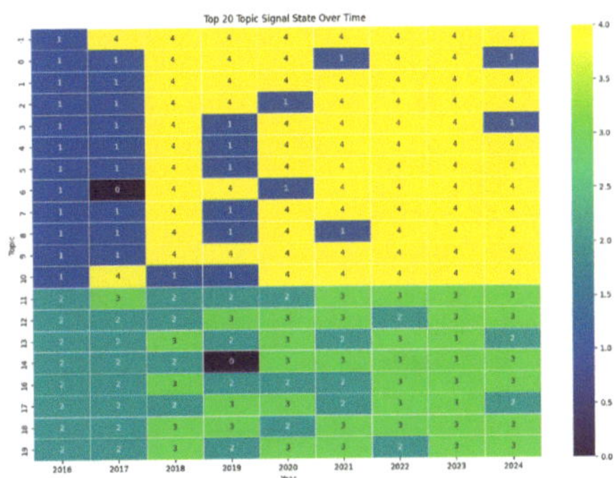

(**i**) Overall trends of EAIs

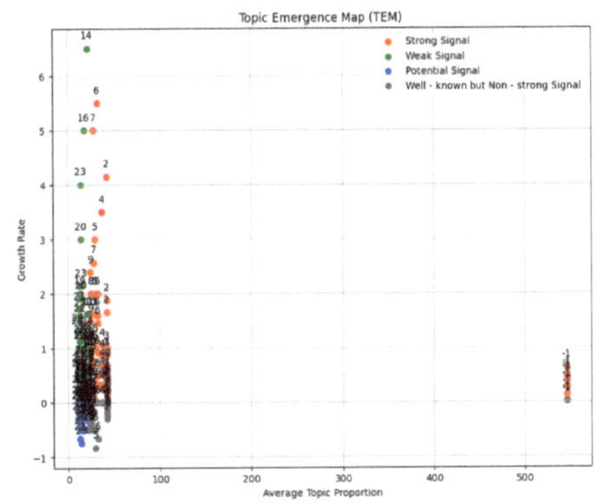

(**j**) Hierarchy clustering of EAIs

Figure 6. This figure shows overall trends for 5 subjects on the left-hand side (Strong Signal (4), Weak Signal (3), Potential Signal (2), Well-known but Non-strong Signal (1), No Signal or Unclassified (0)) and hierarchy clustering on the right-hand side.

Some other examples in the RIS dataset are shown in the table. For instance, Topic 5, '5_terahertz_terahertz wave_vanadium dioxide_resonator', shows that the research on vanadium dioxide resonators for terahertz waves is emerging. Terahertz communication is a key technology for 6G, and vanadium dioxide's unique properties may enable the development of more efficient terahertz devices. Topic 10, '10_holographic_nano brick_micro nano optic_nano optic', indicates the potential of holographic nano-brick-based micro-nano optics in 6G. This could be applied to improve the performance of optical communication components in RIS systems. The heatmap of the TEM for RISs visually represents the development trends of these topics. It allows us to compare the growth rates and average proportions of different topics, facilitating the identification of technologies that are likely to have a significant impact on the future of RIS-related 6G applications. This information can guide companies in the 6G field to allocate resources more effectively and focus on technologies with high-growth potential.

Regarding the ISAC dataset, Topic 15, '15_grating_waveguide_wavelength_coupler', implies that grating waveguide wavelength couplers are emerging as an important technology. These couplers can play a crucial role in optical-based integrated sensing and communication systems, enabling efficient coupling of optical signals. Topic 18, '18_millimeter wave_mimo_hybrid beamforming_mmwave', shows that millimeter-wave MIMO hybrid beamforming technology is gaining momentum. In 6G, MIMO technology combined with hybrid beamforming can enhance communication capacity and signal coverage. The TEM heatmap for ISAC helps visualize the evolution of these topics. It provides a clear picture of which technologies are emerging and which are more established, enabling researchers and industry players to understand the technological landscape better. This understanding is vital for strategic planning, such as setting research priorities and formulating product development strategies in the ISAC domain of 6G.

Table 7. This table shows the weak signals in 5 subjects in patent datasets of 6G.

Datasets	Weak Signals in the Top 20 Topics
NSTs	Topic 17: 17_channel estimation_sparse_millimeter wave channel_deep learning Topic 18: 18_millimeter wave radar_wave radar_doppler_target detection Topic 19: 19_graphene_wave absorber_terahertz wave_substrate layer Topic 20: 20_packaging_packaging structure_millimeter wave chip_wave chip
RISs	Topic 5: 5_terahertz_terahertz wave_vanadium dioxide_resonator Topic 9: 9_display device_pixel_grating_light emitting Topic 10: 10_holographic_nano brick_micro nano optic_nano optic Topic 12: 12_energy efficiency_wireless energy_power_intelligent reflecting surface Topic 13: 13_intelligent reflecting surface_relay_intelligent reflector_communication system Topic 14: 14_supercell_vibration_wave_insulation Topic 15: 15_graphene_absorber_terahertz wave_bias voltage Topic 16: 16_aerial vehicle_unmanned aerial vehicle_unmanned aerial_unmanned Topic 17: 17_positioning_mobile device_reference signal_configuration information Topic 18: 18_wave absorbing_honeycomb_absorber based_metasurface wave Topic 19: 19_vortex wave_orbital angular_vortex beam_oam
ISAC	Topic 15: 15_grating_waveguide_wavelength_coupler Topic 16: 16_offloading_sensor network_wireless sensor network_cluster Topic 17: 17_touch_fingerprint_display device_touch screen Topic 18: 18_millimeter wave_mimo_hybrid beamforming_mmwave Topic 19: 19_underwater_marine_boat_navigation
ISAGSC	Topic 11: 11_frequency offset_clock error_doppler frequency_time frequency Topic 12: 12_laser communication_satellite laser_optical communication_leo satellite Topic 13: 13_conditional_handover command_handover terrestrial_handover terrestrial network Topic 14: 14_switching method_satellite switching_beam switching_satellite base station Topic 15: 15_disclosure relates_system supporting_data transmission rate_pdu Topic 17: 17_constellation_constellation configuration_satellite constellation_design method Topic 18: 18_data transmission method_data transmission_transmission method_transmission Topic 19: 19_edge_edge computing_offloading_ground integrated network
EAIs	Topic 11: 11_transformer_fault_gas_utility Topic 12: 12_syntactic_query_semantic_word vector Topic 13: 13_radar_radar signal_doppler_chirp signal Topic 14: 14_ai model_terminal device_application_channel access Topic 16: 16_voice_speech_audio_microphone Topic 17: 17_ldpc_ldpc code_parity check_channel input Topic 18: 18_amplifier_impedance_transistor_rf signal Topic 19: 19_image quality_image classification_wavelet_image data

In the ISAGSC field, the TEM helps identify emerging trends such as laser communication in satellite systems and edge computing offloading in ground-integrated networks. Weak-signal topics like '12_laser communication_satellite laser_optical communication_leo satellite' and '19_edge_edge computing_offloading_ground integrated network' imply that these areas are emerging as important research directions. This insight is crucial for understanding the future development of 6G integrated space–air–ground–sea communication systems, enabling stakeholders to focus on improving satellite communication and optimizing network computing resources.

Our TEM reveals emerging trends in areas like radar signal processing and AI-based terminal device applications in the EAI dataset. Topics such as '13_radar_radar signal_doppler_chirp signal' and '14_ai model_terminal device_application_channel access' in the weak-signal quadrant suggest that these technologies are evolving. This information is valuable for enhancing the performance of 6G-enhanced air interfaces, guiding research toward better radar-based communication and AI-enabled terminal device capabilities.

Overall, we demonstrate that the DTM framework with TEMs offers valuable insights into emerging technologies in each field. It helps in the early-stage identification of promising technologies, guiding strategic decision-making, resource allocation, and research and development planning in the highly competitive 6G industry.

5. Conclusions

In this study, we aimed to address the challenges of analyzing the technological evolution of 6G through patent data. Given the rapid development and complexity of 6G technologies, traditional topic modeling approaches were insufficient. We proposed a novel dynamic topic modeling framework to overcome these limitations and provide more interpretable alternative tools to analyze emerging trends in the field.

Specifically, we contributed to the usage of LLM-based tools for cost-effective and efficient summarization techniques to pre-process the large volume of patent data. The experimental results show that long summarization improved topic coherence (TC) in most fields, while short summarization enhanced topic diversity (TD). This indicates that different summarization strategies could be applied according to specific analysis requirements, providing a more flexible and efficient way to handle patent data.

Then, we conducted a comparative analysis of various embedding techniques and found no single model was superior across all metrics and datasets. Furthermore, we optimized the hyperparameters of the BERTopic-based framework. Through random searches based on the DBCV value, we significantly improved the performance of the model. The optimized model showed a remarkable increase in TC (about 40% on average) and TD (about 30% on average), enabling more accurate topic representation and a deeper understanding of emerging trends in 6G technology.

Finally, we applied the weak signal analysis method using TEMs in our dynamic topic modeling framework. By analyzing the five 6G patent datasets (NSTs, RISs, ISAC, ISAGSC, and EAIs), we identified numerous emerging technologies. For example, in the NST dataset, topics like '17_channel estimation_sparse_millimeter wave channel_deep learning' and '18_millimeter wave radar_wave radar_doppler_target detection' were recognized as weak signals, indicating emerging research directions in millimeter-wave communication. In the RIS dataset, '5_terahertz_terahertz wave_vanadium dioxide_resonator' and other topics showed potential for future development in terahertz-related and optical-based RIS technologies.

In conclusion, our proposed dynamic topic modeling framework effectively uncovers 6G technological evolution patterns. It provides valuable insights for strategic decision-making in the 6G industry and enables stakeholders to identify emerging technologies, allocate resources rationally, and stay ahead in the highly competitive 6G technology race. Future research could further explore more advanced hyperparameter optimization techniques and expand the scope of patent data analysis to other 6G-related fields to enhance a comprehensive understanding of 6G technological development.

Supplementary Materials: The following supporting information can be downloaded at: https://www.mdpi.com/article/10.3390/app15073783/s1.

Author Contributions: Conceptualization, J.J. and F.Y.; methodology, F.Y.; software, F.Y. and R.D.; validation, J.J. and R.D.; formal analysis, F.Y.; investigation, J.J. and R.D.; resources, J.J.; data curation, J.J.; writing—original draft preparation, J.J.; writing—review and editing, J.J., F.Y., and R.D.; visualization, F.Y.; supervision, J.J.; funding acquisition, F.Y.; project administration, F.Y. and R.D. All authors have read and agreed to the published version of the manuscript.

Funding: This research was funded by National Major Scientific Instruments and Equipments Development Project of National Natural Science Foundation of China, NO. 32327801.

Institutional Review Board Statement: Not applicable.

Informed Consent Statement: Not applicable.

Data Availability Statement: The data presented in this study are available on request from the corresponding author. The data are not publicly available due to the sensitivity of 6G technology competition-related information.

Conflicts of Interest: The authors declare no conflicts of interest.

Abbreviations

The following abbreviations are used in this manuscript:

ADTM	Algorithmic Dynamic Topic Models
BOW	Bag-of-words
BERT	Bidirectional Encoder Representations from Transformers
CBOW	Continuous Bag of Words
DBCV	Density-Based Clustering Validation
DL	Deep Learning
DTM	Dynamic Topic Models
EAI	Enhanced Air Interface
HDBSCAN	Hierarchical Density-Based Spatial Clustering of Applications with Noise
ISAC	Integrated Sensing and Communication
ISAGSC	Integrated Space–Air–Ground–Sea Communication
KEM	Keyword Emergence Maps
KIM	Keyword Issue Maps
LC	Link Clustering
LDA	Latent Dirichlet Allocation
LLMs	Large Language Models
MIMO	Multiple-Input Multiple-Output
NPMI	Normalized Pointwise Mutual Information
NSTs	New Spectrum Technologies
OAM	Orbital Angular Momentum
PDTM	Probabilistic Dynamic Topic Models
PMI	Pointwise Mutual Information
POS	Part-Of-Speech
RIS	Reconfigurable Intelligent Surface
SBERT	Sentence-BERT
TC	Topic Coherence
TD	Topic Diversity
TEMs	Topic Emergence Maps
TF-IDF	Term Frequency–Inverse Document Frequency
UMAP	Uniform Manifold Approximation and Projection

References

1. Pennanen, H.; Hänninen, T.; Tervo, O.; Tölli, A.; Latva-aho, M. 6G: The Intelligent Network of Everything—A Comprehensive Vision, Survey, and Tutorial. *arXiv* **2024**, arXiv:2407.09398.

2. Chowdhury, M.Z.; Shahjalal, M.; Ahmed, S.; Jang, Y.M. 6G Wireless Communication Systems: Applications, Requirements, Technologies, Challenges, and Research Directions. *IEEE Open J. Commun. Soc.* **2020**, *1*, 957–975. [CrossRef]
3. Prasad Tera, S.; Chinthaginjala, R.; Pau, G.; Hoon Kim, T. Toward 6G: An Overview of the next Generation of Intelligent Network Connectivity. *IEEE Access* **2025**, *13*, 925–961. [CrossRef]
4. Na, M.; Lee, J.; Choi, G.; Yu, T.; Choi, J.; Lee, J.; Bahk, S. Operator's Perspective on 6G: 6G Services, Vision, and Spectrum. *IEEE Commun. Mag.* **2024**, *62*, 178–184. [CrossRef]
5. Bazzi, A.; Bomfin, R.; Mezzavilla, M.; Rangan, S.; Rappaport, T.; Chafii, M. Upper Mid-Band Spectrum for 6G: Vision, Opportunity and Challenges. *arXiv* **2025**, arXiv:2502.17914.
6. Lu, S.; Liu, F.; Li, Y.; Zhang, K.; Huang, H.; Zou, J.; Li, X.; Dong, Y.; Dong, F.; Zhu, J.; et al. Integrated Sensing and Communications: Recent Advances and Ten Open Challenges. *IEEE Internet Things J.* **2024**, *11*, 19094–19120. [CrossRef]
7. Meng, S.; Wu, S.; Zhang, J.; Cheng, J.; Zhou, H.; Zhang, Q. Semantics-Empowered Space-Air-Ground-Sea Integrated Network: New Paradigm, Frameworks, and Challenges. *IEEE Commun. Surveys Tuts.* **2025**, *27*, 140–183. [CrossRef]
8. Yuan, Y.; Huang, Y.; Ding, H.; Cui, C.; Wang, Q. *Key Technologies for the 6G Air Interface*; Elsevier: Amsterdam, The Netherlands, 2025; ISBN 978-0-443-33660-7.
9. Hasan, S.R.; Sabuj, S.R.; Hamamura, M.; Hossain, M.A. A Comprehensive Review on Reconfigurable Intelligent Surface for 6G Communications: Overview, Deployment, Control Mechanism, Application, Challenges, and Opportunities. *Wireless Pers. Commun.* **2024**, *139*, 375–429. [CrossRef]
10. Jiang, W.; Han, B.; Habibi, M.A.; Schotten, H.D. The Road towards 6G: A Comprehensive Survey. *IEEE Open J. Commun. Soc.* **2021**, *2*, 334–366. [CrossRef]
11. Trappey, A.J.C.; Huang, A.Z.C.; Chen, N.K.T.; Pa, R.J.S.; Trappey, C.V.; Li, K.A.; Hung, L.P. Transdisciplinary Technology Mining of Advanced 6G Satellite Communication Innovations. In *Advances in Transdisciplinary Engineering*; Moser, B.R., Koomsap, P., Stjepandić, J., Eds.; IOS Press: Amsterdam, The Netherlands, 2022; ISBN 978-1-64368-338-6.
12. Yu, X.; Zhang, B. Obtaining Advantages from Technology Revolution: A Patent Roadmap for Competition Analysis and Strategy Planning. *Technol. Forecast. Soc. Chang.* **2019**, *145*, 273–283. [CrossRef]
13. Krestel, R.; Chikkamath, R.; Hewel, C.; Risch, J. A Survey on Deep Learning for Patent Analysis. *World Pat. Inf.* **2021**, *65*, 102035. [CrossRef]
14. Alghamdi, R.; Alfalqi, K. A Survey of Topic Modeling in Text Mining. *Int. J. Adv. Comput. Sci. Appl.* **2015**, *6*. [CrossRef]
15. Blei, D.M.; Lafferty, J.D. Dynamic Topic Models. In Proceedings of the 23rd International Conference on Machine Learning—ICML '06, Pittsburgh, PA, USA, 25–29 June 2006; ACM Press: Pittsburgh, PA, USA, 2006; pp. 113–120.
16. Latent Dirichlet Allocation (LDA) and Topic Modeling: Models, Applications, a Survey | Multimedia Tools and Applications. Available online: https://link.springer.com/article/10.1007/s11042-018-6894-4 (accessed on 27 January 2025).
17. Egger, R.; Yu, J. A Topic Modeling Comparison Between LDA, NMF, Top2Vec, and BERTopic to Demystify Twitter Posts. *Front. Sociol.* **2022**, *7*, 886498. [CrossRef] [PubMed]
18. Topics over Time | Proceedings of the 12th ACM SIGKDD International Conference on Knowledge Discovery and Data Mining. Available online: https://dl.acm.org/doi/10.1145/1150402.1150450 (accessed on 27 January 2025).
19. Ramos, J. Using TF-IDF to Determine Word Relevance in Document Queries. In Proceedings of the First Instructional Conference on Machine Learning, Virtual, 14 September 2020.
20. Adam, R.; Kogler, M. Tracking the Evolution of Climate Protection Discourse in Austrian Newspapers: A Comparative Study of BERTopic and Dynamic Topic Modeling. *Conf. Proc.* **2023**. [CrossRef]
21. Grootendorst, M. BERTopic: Neural Topic Modeling with a Class-Based TF-IDF Procedure. *arXiv* **2022**, arXiv:2203.05794.
22. Ebadi, A.; Auger, A.; Gauthier, Y. Detecting Emerging Technologies and Their Evolution Using Deep Learning and Weak Signal Analysis. *J. Informetr.* **2022**, *16*, 101344. [CrossRef]
23. Rieger, J.; Jentsch, C.; Rahnenführer, J. RollingLDA: An Update Algorithm of Latent Dirichlet Allocation to Construct Consistent Time Series from Textual Data. In Proceedings of the Findings of the Association for Computational Linguistics: EMNLP 2021, Virtual Event/Punta Cana, Dominican Republic, 16–20 November 2021; Moens, M.-F., Huang, X., Specia, L., Yih, S.W., Eds.; Association for Computational Linguistics: Punta Cana, Dominican Republic, 2021; pp. 2337–2347.
24. Bonino, D.; Ciaramella, A.; Corno, F. Review of the State-of-the-Art in Patent Information and Forthcoming Evolutions in Intelligent Patent Informatics. *World Pat. Inf.* **2010**, *32*, 30–38. [CrossRef]
25. Probabilistic Non-Negative Matrix Factorization and Its Robust Extensions for Topic Modeling | Proceedings of the AAAI Conference on Artificial Intelligence. Available online: https://ojs.aaai.org/index.php/AAAI/article/view/10832 (accessed on 28 January 2025).
26. Arts, S.; Hou, J.; Gomez, J.C. Natural Language Processing to Identify the Creation and Impact of New Technologies in Patent Text: Code, Data, and New Measures. *Res. Policy* **2021**, *50*, 104144. [CrossRef]
27. Obtaining Better Static Word Embeddings Using Contextual Embedding Models—ACL Anthology. Available online: https://aclanthology.org/2021.acl-long.408/ (accessed on 28 January 2025).

28. Mikolov, T.; Sutskever, I.; Chen, K.; Corrado, G.; Dean, J. Distributed Representations of Words and Phrases and Their Compositionality. *arXiv* **2013**, arXiv:1310.4546. Available online: https://arxiv.org/abs/1310.4546 (accessed on 28 January 2025).
29. Le, Q.V.; Mikolov, T. Distributed Representations of Sentences and Documents. *arXiv* **2014**, arXiv:1405.4053.
30. Angelov, D. Top2Vec: Distributed Representations of Topics. *arXiv* **2020**, arXiv:2008.09470.
31. Devlin, J.; Chang, M.-W.; Lee, K.; Toutanova, K. BERT: Pre-Training of Deep Bidirectional Transformers for Language Understanding. *arXiv* **2019**, arXiv:1810.04805.
32. Patent Classification by Fine-Tuning BERT Language Model. *World Pat. Inf.* **2020**, *61*, 101965. [CrossRef]
33. Bekamiri, H.; Hain, D.S.; Jurowetzki, R. PatentSBERTa: A Deep NLP Based Hybrid Model for Patent Distance and Classification Using Augmented SBERT. *Technol. Forecast. Soc. Chang.* **2024**, *206*, 123536. [CrossRef]
34. Rahimi, H.; Naacke, H.; Constantin, C.; Amann, B. ANTM: An Aligned Neural Topic Model for Exploring Evolving Topics. *arXiv* **2023**, arXiv:2302.01501.
35. Integrating Bibliometrics and Roadmapping Methods: A Case of Dye-Sensitized Solar Cell Technology-Based Industry in China—ScienceDirect. Available online: https://www.sciencedirect.com/science/article/pii/S0040162514001644 (accessed on 28 January 2025).
36. Denter, N.; Caferoglu, H.; Moehrle, M.G. Applying Dynamic Topic Modeling for Understanding the Evolution of the RFID Technology. In Proceedings of the 2019 Portland International Conference on Management of Engineering and Technology (PICMET), Portland, OR, USA, 25–29 August 2019; pp. 1–9.
37. Wang, C.; Blei, D.; Heckerman, D. Continuous Time Dynamic Topic Models. *arXiv* **2015**, arXiv:1206.3298.
38. Sharing Clusters Among Related Groups: Hierarchical Dirichlet Processes. Available online: https://papers.nips.cc/paper_files/paper/2004/hash/fb4ab556bc42d6f0ee0f9e24ec4d1af0-Abstract.html (accessed on 28 January 2025).
39. Greene, D.; Cross, J.P. Exploring the Political Agenda of the European Parliament Using a Dynamic Topic Modeling Approach. *arXiv* **2016**, arXiv:1607.03055. Available online: https://arxiv.org/abs/1607.03055 (accessed on 28 January 2025). [CrossRef]
40. Tracking Urban Geo-Topics Based on Dynamic Topic Model—ScienceDirect. Available online: https://www.sciencedirect.com/science/article/pii/S019897151930105X (accessed on 28 January 2025).
41. Frontiers | Using Topic Modeling Methods for Short-Text Data: A Comparative Analysis. Available online: https://www.frontiersin.org/journals/artificial-intelligence/articles/10.3389/frai.2020.00042/full (accessed on 28 January 2025).
42. Boutaleb, A.; Picault, J.; Grosjean, G. BERTrend: Neural Topic Modeling for Emerging Trends Detection. In Proceedings of the Workshop on the Future of Event Detection (FuturED), Miami, FL, USA, 15–16 November 2024; Tetreault, J., Nguyen, T.H., Lamba, H., Hughes, A., Eds.; Association for Computational Linguistics: Miami, FL, USA, 2024; pp. 1–17.
43. Lee, W.H. How to Identify Emerging Research Fields Using Scientometrics: An Example in the Field of Information Security. *Scientometrics* **2008**, *76*, 503–525. [CrossRef]
44. Topic-Linked Innovation Paths in Science and Technology. Available online: https://ideas.repec.org/a/eee/infome/v14y2020i2s175115771930210x.html (accessed on 28 January 2025). [CrossRef]
45. Technological Opportunity Discovery for Technological Convergence Based on the Prediction of Technology Knowledge Flow in a Citation Network—ScienceDirect. Available online: https://www.sciencedirect.com/science/article/pii/S1751157718300907 (accessed on 28 January 2025). [CrossRef]
46. Ansoff, H.I. Managing Strategic Surprise by Response to Weak Signals. *Calif. Manage. Rev.* **1975**, *18*, 21–33. [CrossRef]
47. Xu, H.; Luo, R.; Winnink, J.; Wang, C.; Elahi, E. A Methodology for Identifying Breakthrough Topics Using Structural Entropy. *Inf. Process. Manag. Int. J.* **2022**, *59*, 102862. Available online: https://dl.acm.org/doi/10.1016/j.ipm.2021.102862 (accessed on 28 January 2025).
48. Yoon, J. Detecting Weak Signals for Long-Term Business Opportunities Using Text Mining of Web News. *Expert Syst. Appl. Int. J.* **2012**, *39*, 12543–12550. Available online: https://dl.acm.org/doi/10.1016/j.eswa.2012.04.059 (accessed on 28 January 2025).
49. Maitre, J.; Menard, M.; Chiron, G.; Bouju, A. Détection de Signaux Faibles Dans Des Masses de Données Faiblement Structurées. *Rech. Inf. Doc. Web Sémant.* **2019**, *3*. [CrossRef]
50. El Akrouchi, M.; Benbrahim, H.; Kassou, I. End-to-End LDA-Based Automatic Weak Signal Detection in Web News. *Knowledge-Based Syst.* **2021**, *212*, 106650. [CrossRef]
51. Ebadi, A.; Auger, A.; Gauthier, Y. WISDOM: An AI-Powered Framework for Emerging Research Detection Using Weak Signal Analysis and Advanced Topic Modeling. *arXiv* **2024**, arXiv:2409.15340.
52. incoPat Reliable and User-Friendly Global Patent Database Patent Search Patent Analyze Intellectual Property incoPat. Available online: https://www.incopat.com/ (accessed on 28 January 2025).
53. Khandelwal, T. Investigating the Impact of Text Summarization on Topic Modeling. *arXiv* **2024**, arXiv:2410.09063.
54. Salton, G.; Buckley, C. Term-Weighting Approaches in Automatic Text Retrieval. *Inf. Process. Manag.* **1988**, *24*, 513–523. [CrossRef]
55. Baevski, A.; Hsu, W.-N.; Xu, Q.; Babu, A.; Gu, J.; Auli, M. Data2vec: A General Framework for Self-Supervised Learning in Speech, Vision and Language. *arXiv* **2022**, arXiv:2202.03555.
56. Reimers, N.; Gurevych, I. Sentence-BERT: Sentence Embeddings Using Siamese BERT-Networks. *arXiv* **2019**, arXiv:1908.10084.

57. Campello, R.J.G.B.; Moulavi, D.; Sander, J. Density-Based Clustering Based on Hierarchical Density Estimates. In *Proceedings of the Advances in Knowledge Discovery and Data Mining*; Pei, J., Tseng, V.S., Cao, L., Motoda, H., Xu, G., Eds.; Springer: Berlin/Heidelberg, Germany, 2013; pp. 160–172.
58. Newman, D.; Lau, J.H.; Grieser, K.; Baldwin, T. Automatic Evaluation of Topic Coherence. In Proceedings of the Human Language Technologies: The 2010 Annual Conference of the North American Chapter of the Association for Computational Linguistics, Los Angeles, CA, USA, 1–6 June 2010; Kaplan, R., Burstein, J., Harper, M., Penn, G., Eds.; Association for Computational Linguistics: Los Angeles, CA, USA, 2010; pp. 100–108.
59. Röder, M.; Both, A.; Hinneburg, A. Exploring the Space of Topic Coherence Measures. In Proceedings of the Eighth ACM International Conference on Web Search and Data Mining, New York, NY, USA, 2–6 February 2015; Association for Computing Machinery: New York, NY, USA, 2015; pp. 399–408.
60. Bouma, G. Normalized (Pointwise) Mutual Information in Collocation Extraction. In Proceedings of the Biennial GSCL Conference 2009, Bergen, Norway, 17–20 September 2009.
61. Church, K.W.; Hanks, P. Word Association Norms, Mutual Information, and Lexicography. *Comput. Linguist.* **1990**, *16*, 22–29.
62. Hashimoto, T.; Shepard, D.L.; Kuboyama, T.; Shin, K.; Kobayashi, R.; Uno, T. Analyzing Temporal Patterns of Topic Diversity Using Graph Clustering. *J. Supercomput.* **2021**, *77*, 4375–4388. [CrossRef]
63. Beltagy, I.; Lo, K.; Cohan, A. SciBERT: A Pretrained Language Model for Scientific Text. *arXiv* **2019**, arXiv:1903.10676.
64. Liu, Y.; Ott, M.; Goyal, N.; Du, J.; Joshi, M.; Chen, D.; Levy, O.; Lewis, M.; Zettlemoyer, L.; Stoyanov, V. RoBERTa: A Robustly Optimized BERT Pretraining Approach. *arXiv* **2019**, arXiv:1907.11692.

applied sciences

MDPI

Article

Video Temporal Grounding with Multi-Model Collaborative Learning

Yun Tian [1], Xiaobo Guo [1,*], Jinsong Wang [1], Bin Li [2] and Shoujun Zhou [2]

[1] School of Optoelectronic Engineering, Changchun University of Science and Technology, Changchun 130022, China; tianyun@mails.cust.edu.cn (Y.T.); wang.jinsong@hotmail.com (J.W.)
[2] Shenzhen Institute of Advanced Technology, Chinese Academy of Sciences, Shenzhen 518055, China; b.li2@siat.ac.cn (B.L.); sj.zhou@siat.ac.cn (S.Z.)
* Correspondence: guoxiaobo1967@126.com

Abstract: Given an untrimmed video and a natural language query, the video temporal grounding task aims to accurately locate the target segment within the video. Functioning as a critical conduit between computer vision and natural language processing, this task holds profound importance in advancing video comprehension. Current research predominantly centers on enhancing the performance of individual models, thereby overlooking the extensive possibilities afforded by multi-model synergy. While knowledge flow methods have been adopted for multi-model and cross-modal collaborative learning, several critical concerns persist, including the unidirectional transfer of knowledge, low-quality pseudo-label generation, and gradient conflicts inherent in cooperative training. To address these issues, this research proposes a Multi-Model Collaborative Learning (MMCL) framework. By incorporating a bidirectional knowledge transfer paradigm, the MMCL framework empowers models to engage in collaborative learning through the interchange of pseudo-labels. Concurrently, the mechanism for generating pseudo-labels is optimized using the CLIP model's prior knowledge, bolstering both the accuracy and coherence of these labels while efficiently discarding extraneous temporal fragments. The framework also integrates an iterative training algorithm for multi-model collaboration, mitigating gradient conflicts through alternate optimization and achieving a dynamic balance between collaborative and independent learning. Empirical evaluations across multiple benchmark datasets indicate that the MMCL framework markedly elevates the performance of video temporal grounding models, exceeding existing state-of-the-art approaches in terms of mIoU and Rank@1. Concurrently, the framework accommodates both homogeneous and heterogeneous model configurations, demonstrating its broad versatility and adaptability. This investigation furnishes an effective avenue for multi-model collaborative learning in video temporal grounding, bolstering efficient knowledge dissemination and charting novel pathways in the domain of video comprehension.

Keywords: video temporal grounding; collaborative learning; pseudo-label; iterative training

1. Introduction

Video Temporal Grounding (VTG) seeks to accurately pinpoint the corresponding time segment in an untrimmed video based on natural language queries [1]. As a pivotal task at the intersection of computer vision and natural language processing, VTG lays the groundwork for a variety of downstream video comprehension tasks, such as video dialogue systems [2], video relationship detection [3–5], and video question answering systems [6–8]. The majority of traditional VTG strategies concentrate on refining a single model, typified by the Span-based Model and the 2D-Map Model shown in Figure 1, both of which display distinct advantages and limitations in particular contexts [9]. The Span-based Model explicitly forecasts the probabilistic distribution of a target interval's start and end points, showcasing commendable capabilities in delineating precise temporal boundaries. Nevertheless, given its dependence on localized feature extraction, it exhibits a diminished aptitude for holistic semantic context modeling, thereby restricting its effectiveness in handling semantically intricate videos [10,11]. In comparison, the 2D-Map Model devises a two-dimensional temporal mapping, partitioning the entire video span into a 2D matrix in which every element corresponds to a prospective temporal clip. Operating within a two-dimensional framework, this model maximizes the score region, thereby adeptly capturing the contextual interplay among temporal segments. However, the expansive search space for candidate clips renders its computational overhead substantial, ultimately undermining boundary prediction precision [12–14].

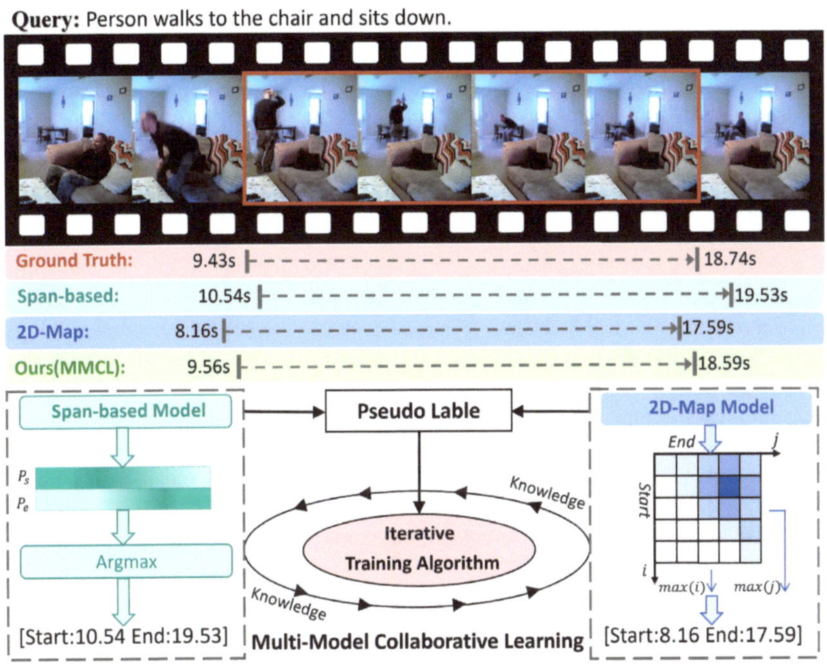

Figure 1. An illustration of localizing a temporal moment in an untrimmed video using a language query. The 2D-MAP Model and the Span-based Model employ pseudo-labels to facilitate knowledge transfer and achieve collaborative learning across models. Leveraging an iterative training algorithm, they strike a dynamic balance between collaborative learning and independent optimization.

Despite the impressive performance of these models in specific settings, their independent optimization strategies fail to fully exploit the complementary nature among models, disregarding the potential for collaborative learning and thereby limiting further performance enhancement. In recent years, knowledge flow, which encompasses knowledge distillation and transfer, has increasingly become a critical strategy for achieving multi-model cooperative optimization in VTG endeavors. Through the supervised transfer of extensive knowledge from large-scale models to more compact ones, knowledge distillation empowers lightweight models to attain robust performance under constrained computational budgets [9,15,16]. By leveraging cross-modal alignment and data exchange, knowledge transfer fortifies the synergy between visual and textual representations, facilitating effective multimodal co-modeling [17,18]. Nevertheless, existing work often restricts knowledge flow to a unidirectional paradigm, such as transferring insights from larger models to smaller counterparts or employing simplistic cross-modal alignments. Such approaches generate pseudo-labels of limited quality and are prone to gradient conflicts in multi-model collaboration [19,20], thus failing to fully harness the potential of bidirectional cooperative learning and dynamic knowledge sharing. These shortcomings have emerged as a key bottleneck restricting further advancements in VTG performance.

To solve the aforementioned deficiencies, this study introduces a Multi-Model Collaborative Learning (MMCL) framework, illustrated in Figure 1. Within this framework, a bidirectional collaborative learning paradigm is instituted across multiple models, yielding significant enhancements in the performance of video temporal grounding. Concretely, to surmount the constraints of one-way knowledge transfer, the MMCL architecture adopts a bidirectional knowledge transmission scheme. Through an equitable exchange of pseudo-labels across different models, predictions from Model A furnish pseudo-supervision for Model B, while Model B's output similarly informs Model A. This bidirectional collaboration mechanism fully leverages the complementary learning capacities among models, compensates for the limitations of one-way knowledge flow, and further capitalizes on each model's strengths in boundary prediction and context modeling. To bolster the fidelity of pseudo-labels, the MMCL framework introduces a CLIP-guided pseudo-label generator. Leveraging the CLIP [21] model's cross-modal priors, this module calculates the similarity between the video and the query text to provide precise temporal constraints for the pseudo-label generator, thereby reducing noise while enhancing temporal accuracy and semantic consistency of the pseudo-labels. Moreover, to counteract gradient clashes within multi-model collaborative learning, the MMCL framework embeds an iterative training algorithm. During each optimization cycle, one model's parameters remain static while the other model's supervision and transfer objectives are refined. This alternating optimization tactic substantially alleviates gradient contention among models and ensures their respective optimization autonomy. Through adaptive calibration of consistency thresholds and transfer loss coefficients, the framework achieves a nuanced balance between cooperative learning and autonomous optimization, culminating in superior overall performance.

Experimental results show that the proposed MMCL framework exhibits pronounced strengths and robust versatility in video temporal grounding. Its inherent adaptability makes the MMCL framework suitable for both heterogeneous and homogeneous model collaboration contexts, where it consistently yields appreciable performance gains. From a theoretical perspective, this study broadens the applicability of knowledge transfer and collaborative learning, addressing limitations in pseudo-label quality and model collaboration mechanisms, while offering an efficient and flexible solution for video temporal grounding. The key contributions of this work are enumerated as follows:

- We propose a multi-model collaborative learning framework employing a bidirectional knowledge transfer scheme to foster cooperative learning across distinct models, thus transcending the confines of isolated model optimization and amplifying cross-model complementarity.
- We design a CLIP-guided pseudo-label generator that harnesses CLIP's cross-modal priors to optimize pseudo-label generation, improving their quality and robustness, and bolstering the accuracy of knowledge transfer.
- We present an iterative multi-model training algorithm that alternately freezes and optimizes model parameters to effectively mitigate gradient conflicts, safeguard the balance between cooperative and independent optimization, and further boost training stability and model performance.

- Experimental findings reveal that our method effectively enhance the performance of various baseline methods for video temporal grounding. By incorporating both the VSLNet [22] and 2D-TAN [23] baseline models, performance on multiple datasets surpasses current mainstream approaches.

2. Related Work

2.1. Video Temporal Grounding

Video Temporal Grounding (VTG) involves aligning temporal video segments with natural language queries, and existing methods predominantly fall into two categories: Span-based Models and 2D-MAP Models. Span-based Models, such as VSLNet [22], frame the VTG task as a span-based question-answering problem, where video segments serve as context, the target temporal fragment as the answer span, and the linguistic query as the question. These models excel at accurately detecting temporal boundaries and efficiently localizing target segments through boundary regression techniques. However, their focus on explicit boundary prediction often limits their ability to capture broader semantic contexts and handle long-duration interactions, making them less effective in complex video scenarios [10,11]. In contrast, 2D-MAP Models encode all potential start-end intervals as elements of a two-dimensional matrix, refining high-scoring regions to locate the desired temporal segment [9]. The 2D-TAN model [23], for example, constructs a temporal map to model contextual interactions among video segments, offering robust capabilities in modeling context and segment dependencies [24]. Despite their strengths, these models may struggle with boundary prediction accuracy in densely overlapping temporal segments, particularly under intricate temporal distributions [12–14]. While both approaches offer distinct advantages, their independent optimization limits the potential of leveraging their complementary strengths, highlighting the need for a collaborative learning framework like MMCL to integrate these methodologies effectively.

2.2. Knowledge Flow in Video Temporal Grounding

The utilization of knowledge flow in multimodal tasks is gaining momentum, predicated on the concept of disseminating knowledge across different models or modalities to augment collective model capability [25]. Present investigations can be bifurcated into two principal streams: knowledge distillation and knowledge transfer. Via knowledge distillation, the expertise of a large model is conferred upon a more compact model, ensuring the latter maintains commendable performance while remaining computationally efficient [26]. By integrating both local and global modality information and applying distillation techniques, Bao et al. improved the feature representation of video understanding models, thereby achieving greater accuracy in temporal localization tasks [16]. Other works leverage the robust representational capacity of large language models (LLMs) to generate weakly supervised pseudo-labels via distillation, thereby offering a novel optimization pathway for VTG tasks [15]. Furthermore, Liang et al. introduced a distillation approach employing multiple teacher models to collectively steer the student model's learning, significantly boosting temporal localization accuracy for VTG [9]. Diverging from the principles of knowledge distillation, knowledge transfer focuses on facilitating cross-modal information exchanges that foster cooperative learning among distinct modalities. Weng et al. proposed a vision-prompted knowledge transfer method that integrates visual and textual modality features, significantly enhancing multimodal reasoning in video question answering tasks [17]. Wu et al. devised a dynamic knowledge transfer mechanism that enables cross-modal knowledge sharing between visual and textual predictors, effectively bolstering temporal localization and answer generation in video question answering tasks [18]. Although existing knowledge flow methods moderately improve model performance, most studies are confined to unidirectional knowledge transfer (e.g., from large to small models) or cross-modal knowledge migration, failing to fully explore the potential of bidirectional cooperative learning and thus limiting the depth and breadth of knowledge sharing. Additionally, the phenomenon of gradient contention within multi-model collaborative learning undercuts the advantages of coordinated optimization, thus imposing further limitations on knowledge-sharing efficacy and holistic performance improvements.

2.3. Pseudo-Label in Video Temporal Grounding

Pseudo-labels are extensively utilized in video temporal grounding tasks, fundamentally relying on model-generated predictions to form surrogate supervision signals. Such an approach bolsters performance in weakly or semi-supervised settings, while likewise demonstrating considerable viability in low-resource or unsupervised domains [27]. Xu et al. proposed a pseudo-label generation method based on predictive confidence, where high-confidence time segments are filtered for training via a confidence threshold [28]. Nevertheless, this approach remains vulnerable to threshold selection, risking the exclusion of valuable segments or the inclusion of suboptimal ones [29]. An alternative tactic is to utilize the model's predicted temporal segments as pseudo-labels outright, for instance, leveraging initial predictions as pseudo supervision in weakly supervised video localization [30]. Yet the fidelity of such pseudo-labels is wholly determined by the initial model's predictive prowess, impeding effective mitigation of noisy segments [30]. While these methods may excel in particular scenarios, pseudo-label quality generally hinges on the accuracy of the underlying model predictions [31]. Inferior pseudo-labels can erode the efficacy of model training and inject noise into the learning procedure, thereby impeding potential gains in performance [32]. Moreover, existing methods lack explicit constraints on the pseudo-label generation range, making it difficult for the model to filter out irrelevant or low-confidence segments.

3. Methodology

In this study, we introduce an MMCL framework tailored for video temporal grounding. The overall architecture of MMCL is illustrated in Figure 2.

3.1. Problem Formulation

The objective of video temporal grounding is to identify the optimal time segment M, consistent with the query's semantics, given an untrimmed video V of duration L and a natural language query Q. In this task, the video V is denoted by a frame sequence $V = \{f_i\}_{i=1}^{T}$, where f_i represents the i-th frame of the video and T is the total number of frames; the natural language query $Q = \{q_j\}_{j=1}^{W}$ comprises the j-th word of the query, with W being the total number of words. Presently, mainstream approaches to video temporal grounding typically employ a multimodal fusion paradigm, mapping both the video V and the query Q into a unified shared feature space X. Within the feature space X, the model learns the probability distribution of the time range, predicting the start and end times $[t_s, t_e]$ of the target segment to ensure high semantic consistency with the query.

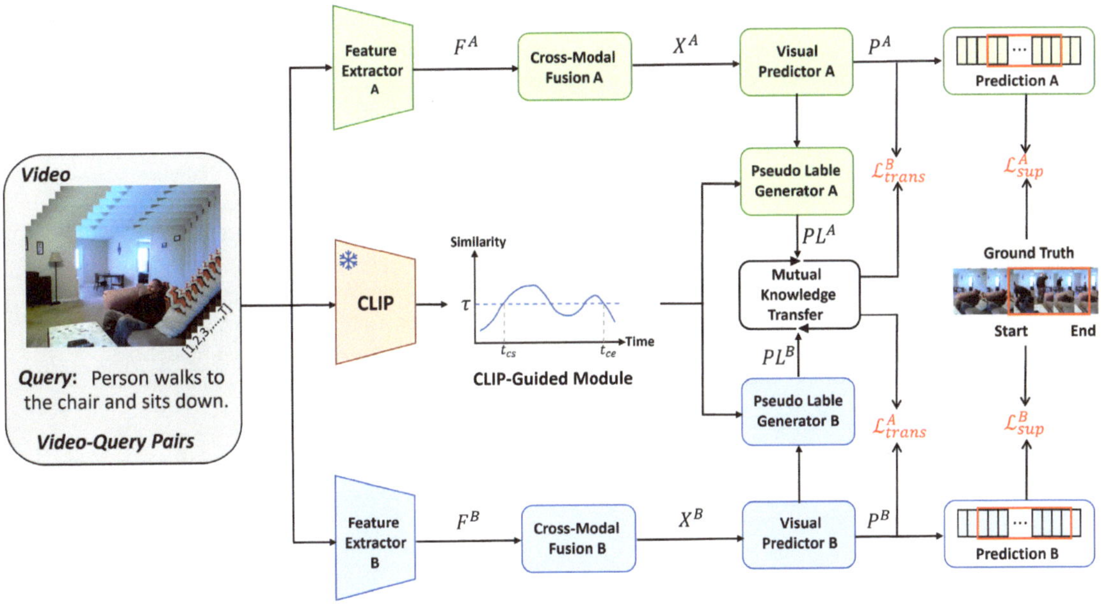

Figure 2. Overview of the proposed Multi-Model Collaborative Learning (MMCL) architecture. The framework comprises: (1) two general models; (2) a CLIP-guided module; (3) a pseudo-label generator; (4) an iterative training algorithm.

3.2. General Model

The base architecture of the MMCL framework encompasses a feature extractor, a cross-modal fusion module, and a visual predictor. To address various model requirements under the MMCL framework, the basic model unifies the structure of video temporal grounding architectures, enabling effective multi-model cooperation. The feature extractor obtains features from the input video V and the natural language query Q, encoding both into high-dimensional representations used by the subsequent cross-modal fusion component to model their semantic interconnections. The video is partitioned into N equally spaced segments, each of duration $\Delta t = \frac{L}{N}$. A pretrained visual model (e.g., 3D ConvNet [33] or CNN [34]) extracts the visual features for each segment, denoted as follows:

$$F_V^k = \text{FeatureExtractor}_V(V), \quad F_V^k \in \mathbb{R}^{N \times d_v} \tag{1}$$

Here, d_v is the dimensionality of the video features, and k signifies the current model. The query is processed by a pretrained linguistic model (e.g., GloVe [35] or BERT [36]) to obtain features, denoted by the following:

$$F_Q^k = \text{FeatureExtractor}_Q(Q), \quad F_Q^k \in \mathbb{R}^{M \times d_q} \tag{2}$$

In this case, d_q represents the dimensional size of the query feature vector. The aforementioned extracted features are subsequently passed to the Cross-Modal Fusion module. The Cross-Modal Fusion module combines the video features F_V^k and the query features F_Q^k to generate the multimodal fused representation X^k, formally given by the following:

$$X^k = \text{Fusion}_k(F_V^k, F_Q^k), \quad X^k \in \mathbb{R}^{N \times d_x} \tag{3}$$

In this formulation, d_x signifies the dimensional scale of the fused feature space. The Cross-Modal Fusion module effectively captures the interactive information between video segments and the query semantics, providing high-quality multimodal representations to the visual prediction module. Under the MMCL framework, each model maintains its native multimodal processing approach while leveraging pseudo-labels to learn from the strengths of the other model. MMCL acts as a flexible collaborative layer that supports diverse architectural choices and fosters model synergy. By using pseudo-labels as soft supervision signals, MMCL facilitates knowledge transfer without altering the inherent multimodal interaction strategies of each model, allowing, for example, the 2D-TAN model to learn fine-grained boundary prediction from VSLNet in fast-changing scenarios, and vice versa. The visual predictor estimates the target time range P_k from the fused features X_k, with the output format varying by model architecture. Span-based models yield 1D probability distributions for the start and end time points, expressed as follows:

$$P_s^k, P_e^k = \text{Predictor}_k(X^k), \quad P_s^k, P_e^k \in \mathbb{R}^N \tag{4}$$

Meanwhile, 2D-MAP-based models output a two-dimensional temporal probability matrix, represented by the following:

$$P_{i,j}^k = \text{Predictor}_k(X^k), \quad P_{i,j}^k \in \mathbb{R}^{N \times N} \tag{5}$$

One dimension i of the matrix corresponds to the start boundary, and the other dimension j to the end boundary, with $i < j$. Each element $P_{i,j}$ in the matrix represents the probability that the video segment begins at index i and ends at index j.

3.3. *Pseudo-Label Generator*

3.3.1. CLIP-Guided Module

The CLIP-Guided Module leverages the cross-modal alignment capability of the CLIP model between video frames and textual queries, providing a coarse-grained temporal constraint for pseudo-label generation to ensure semantic consistency. The overall process can be broken down into three distinct stages:

(1) Semantic Vector Extraction

The input video V is uniformly sampled into n_f frames across its total length L, represented by $\{f_i\}_{i=1}^{n_f}$. Every frame is paired with a timestamp $\{t_i\}_{i=1}^{n_f}$, where $t_i = \frac{i}{n_f} \cdot L$. For each frame f, the CLIP model's visual encoder yields its semantic vector v_i, formulated as follows:

$$v_i = \text{CLIP}_{\text{image}}(f_i), \quad i = 1, 2, \ldots, n_f \tag{6}$$

Concurrently, the query Q is passed through CLIP's text encoder to obtain its textual representation:

$$u_i = \text{CLIP}_{\text{text}}(Q) \tag{7}$$

(2) Computing Similarities

To gauge the semantic congruence between v_i and u_i, we calculate their cosine similarity. The calculation is as follows:

$$S_i = \text{cosine\_similarity}(v_i, u_i) = \frac{v_i \cdot u_i}{\|v_i\| \|u_i\|}, \quad i = 1, 2, \ldots, n_f \tag{8}$$

In this formulation, $S_i \in [-1, 1]$, with higher values signifying greater semantic harmony between Q and frame f_i. The sequence of similarity scores for the n_f frames is denoted by $S = \{S_1, S_2, \ldots, S_{n_f}\}$.

(3) Generating the Constrained Interval

Drawing on the similarity sequence S, a threshold τ is utilized to filter relevant frames by their time span. First, all frames with $S_i \geq \tau$ are identified, and their minimum and maximum indices, i_{min} and i_{max}, are determined according to Equation (9).

$$i_{\min} = \min\{i \mid S_i \geq \tau\}, \quad i_{\max} = \max\{i \mid S_i \geq \tau\} \tag{9}$$

Subsequently, these frame indices are mapped to actual temporal boundaries: $t_{cs} = t_{i_{\min}}; \quad t_{ce} = t_{i_{\max}}$. Finally, $[t_{cs}, t_{ce}]$ denotes the coarse-grained temporal interval guided by the CLIP model, which constrains pseudo-label generation.

3.3.2. Pseudo-Labels Generation

In the process of pseudo-label generation, we utilize the coarse-grained temporal interval $[t_{cs}, t_{ce}]$ provided by the CLIP model and apply normalization to constrain the range of pseudo-label creation, thereby ensuring semantic consistency and producing pseudo-labels $[\hat{t}_s, \hat{t}_e]$ that align with semantic priors. The generation of pseudo-labels depends on the prediction format of different models. Below, we separately describe the pseudo-label generation processes for Span-based models and 2D-MAP models.

(1) Pseudo-Labels Generation for Span-based Models

As stated in Section 3.2, a Span-based model (such as VSLNet) produces probability distributions $P_s(t)$ and $P_e(t)$ corresponding to the start and end times, respectively. The original probability distributions are normalized over the entire video span $[1, N]$, which contains a substantial number of query-irrelevant time segments. To enhance semantic consistency, it is necessary to constrain and re-normalize these distributions based on the temporal window $[t_{cs}, t_{ce}]$ provided by CLIP. Following restriction of these probability distributions to the selected interval, the normalized forms $\hat{P}_s(t)$ and $\hat{P}_e(t)$ are specified as follows:

$$\hat{P}_s(t) = \begin{cases} \dfrac{P_s(t)}{\sum_{t=t_{cs}}^{t_{ce}} P_s(t)}, & t \in [t_{cs}, t_{ce}], \\ 0, & \text{otherwise.} \end{cases} \tag{10}$$

$$\hat{P}_e(t) = \begin{cases} \dfrac{P_e(t)}{\sum_{t=t_{cs}}^{t_{ce}} P_e(t)}, & t \in [t_{cs}, t_{ce}], \\ 0, & \text{otherwise.} \end{cases} \tag{11}$$

The normalization step guarantees that the distributions remain confined to semantically pertinent intervals, thereby narrowing the pseudo-label search domain and fortifying semantic coherence. Upon finalizing the normalization, the pseudo-label boundaries $\hat{t}_s$ and $\hat{t}_e$ emerge by identifying the indices that maximize the normalized probability distributions, formally expressed as follows:

$$\text{index}_s = \arg\max \hat{P}_s(t), \quad \text{index}_e = \arg\max \hat{P}_e(t) \tag{12}$$

Subsequently, the indices are mapped to pseudo-labels PL^k by Equation (13).

$$PL^k = [\hat{t}_s = \text{index}_s \times \Delta t, \hat{t}_e = \text{index}_e \times \Delta t] \tag{13}$$

(2) Pseudo-Labels Generation for 2D-MAP Models

Under the 2D-MAP model, the pseudo-labels derive from a 2D temporal probability matrix $P_{i,j}$, with i and j designating the start and end indices, respectively, and adhering to $i \leq j$. For $P_{i,j}$, elements within the CLIP-guided temporal window are retained, while probabilities outside this range are set to zero. The remaining elements are then normalized to ensure the sum is one. The definition of the normalized probability matrix $\hat{P}_{i,j}$ is as follows:

$$\hat{P}_{i,j} = \begin{cases} \frac{P_{i,j}}{\sum_{i=t_{cs}}^{t_{ce}} \sum_{j=i}^{t_{ce}} P_{i,j}}, & i,j \in [t_{cs}, t_{ce}] \text{ and } i \leq j, \\ 0, & \text{otherwise.} \end{cases} \quad (14)$$

Subsequently, the highest-probability entry in $\hat{P}_{i,j}$ pinpoints the start and end indices of the pseudo-labels:

$$(\text{index}_i, \text{index}_j) = \arg\max \hat{P}_{i,j} \quad (15)$$

Lastly, Equation (16) translates these indices into the pseudo-labels PL_k.

$$PL^k = [\hat{t}_s = \text{index}_i \times \Delta t, \hat{t}_e = (\text{index}_j + 1) \times \Delta t] \quad (16)$$

By following the procedure above, we leverage CLIP's semantic guidance to generate high-quality pseudo-labels, thus offering reliable supervisory signals for training.

3.4. Mutual Knowledge Transfer

At the heart of the MMCL framework is the harnessing of inter-model knowledge exchange and cooperative optimization, capitalizing on each model's complementary capabilities to boost collective outcomes. Within this framework, pseudo-label generation and loss function design constitute the essential elements for enabling collaborative learning. Pseudo-label generation (detailed in Section 3.3) furnishes high-quality supervisory signals for cross-model knowledge migration, while the loss function employs a suitable optimization strategy to effectively merge single-model learning with inter-model collaboration. Accordingly, we construct the comprehensive loss function as a weighted amalgamation of supervision loss and transfer loss, satisfying both each model's need for accurate learning and the bidirectional knowledge exchange among multiple models. In the supervised learning stage, the supervision loss steers Models A and B to refine their performance using genuine labels as guidance. Formally, it is specified as follows:

$$L^A_{\text{sup}} = \mathcal{L}(P^A, GT), \quad L^B_{\text{sup}} = \mathcal{L}(P^B, GT) \quad (17)$$

Here, P^A and P^B denote the predictions of Model A and Model B, respectively; GT stands for ground truth; and $\mathcal{L}$ refers to the particular form of the loss function, which varies depending on the model category. The role of the supervision loss is to align P^A and P^B more tightly with GT, providing direct supervised feedback to each model. In the knowledge transfer phase, the transfer loss drives collaborative learning between models by optimizing the pseudo-labels produced by the counterpart model, and leverages those pseudo-labels as indirect supervision signals to achieve cross-model knowledge transfer. Specifically, Model A's transfer loss L^A_{trans} uses the pseudo-labels PL^B produced by Model B, aligning them with A's predictions P^A to encourage Model A to learn B's predictive characteristics. Analogously, Model B's transfer loss L^B_{trans} imposes knowledge constraints on Model B via PL^A, nudging its optimization objective toward Model A's predictions. The transfer loss is defined as follows:

$$L^A_{\text{trans}} = \mathcal{L}(P^A, PL^B), \quad L^B_{\text{trans}} = \mathcal{L}(P^B, PL^A) \quad (18)$$

Building upon the aforementioned supervision and transfer objectives, the MMCL framework accomplishes multi-model cooperative learning through a weighted synthesis of these losses. The aggregate loss function is given by the following:

$$\text{LOSS} = L_{\text{sup}}^{A} + L_{\text{sup}}^{B} + \alpha L_{\text{trans}}^{A} + \beta L_{\text{trans}}^{B} \tag{19}$$

Here, α and β serve as hyperparameters governing the balance between transfer losses. With this scheme, every model benefits from solitary supervision-driven optimization toward ground-truth labels and, in parallel, participates in inter-model knowledge exchange facilitated by the transfer losses.

3.5. Iterative Training Algorithm

Within the MMCL framework, the core aim of the loss optimization strategy is to foster coherent predictions among models, yet preserve each model's autonomy in its individual optimization processes. To accomplish this, we introduce an iterative training algorithm that blends consistency evaluation with dynamic loss tuning, thereby enabling inter-model knowledge exchange and collective enhancement. In this section, we elaborate on the loss function's optimization strategy from three perspectives: consistency assessment, dynamic loss adjustment, and algorithmic implementation.

3.5.1. Consistency Assessment

Within the realm of multi-model collaborative learning, the temporal intervals forecasted by various models can at times diverge, resulting in inconsistency. To assess and quantify this inconsistency, we introduce the Intersection over Union (IoU) as a measure of consistency. Concretely, IoU is calculated by dividing the intersection of the two predicted time spans by their union, expressed as follows:

$$\text{IoU}(A,B) = \frac{\max(0, \min(t_e^A, t_e^B) - \max(t_s^A, t_s^B))}{(t_e^A - t_s^A) + (t_e^B - t_s^B) - \max(0, \min(t_e^A, t_e^B) - \max(t_s^A, t_s^B))} \tag{20}$$

In this context, $[t_s^A, t_e^A]$ and $[t_s^B, t_e^B]$, respectively, signify the predicted time intervals for Model A and Model B. On the basis of the IoU value, we define an IoU threshold λ to gauge the degree of concordance between the models' predictions. When IoU $> \lambda$, the alignment between the two models' predictions is relatively strong, prompting a decrease in the transfer loss weight. Conversely, if IoU $\leq \lambda$, the predictions are deemed inconsistent, warranting an increase in the transfer loss weight so as to guide both models' results toward consistency. By quantifying the degree of consistency, the MMCL framework can dynamically adjust inter-model learning intensity, ensuring the effectiveness of collaborative learning. Consistency evaluation offers a dynamic adjustment signal that directly influences the adaptive loss adjustment procedure (see Section 3.5.2), guiding transfer loss weighting and providing a basis for iterative training.

3.5.2. Dynamic Loss Adjustment

To balance the independent optimization of each model with the consistency of cross-model predictions in collaborative learning, we devise a dynamic loss adjustment strategy founded on IoU. Building on top of the supervision and transfer losses, we incorporate a dynamic weighting factor ω (refer to Equation (21)) to regulate the influence of transfer loss on the total loss. The definition of the dynamic factor ω is as follows:

$$\omega = \begin{cases} 1 - \text{IoU}(A,B), & \text{if IoU} \leq \lambda, \\ \gamma \cdot (1 - \text{IoU}(A,B)), & \text{if IoU} > \lambda \end{cases} \tag{21}$$

Here, $\gamma \in (0,1)$ is a parameter that governs the attenuation of the transfer loss under high consistency conditions. The total loss function after dynamic adjustment can be expressed as follows:

$$\text{LOSS} = L_{\text{sup}}^{A} + L_{\text{sup}}^{B} + \alpha \cdot \omega \cdot L_{\text{trans}}^{A} + \beta \cdot \omega \cdot L_{\text{trans}}^{B} \quad (22)$$

In this expression, α and β are the intrinsic weights for the transfer loss, and the factor ω modulates the strength of cross-model transfer in accordance with IoU. Under conditions of strong predictive alignment, the framework grants each model greater latitude for autonomous optimization, whereas lower alignment triggers a heavier emphasis on cross-model knowledge exchange, thereby bolstering model concordance. Leveraging this strategy, the MMCL framework deftly orchestrates the magnitudes of both supervision and transfer losses, securing an optimal trade-off between inter-model knowledge sharing and standalone optimization.

3.5.3. Algorithmic Realization

In the course of knowledge dissemination, inter-model gradients can impede one another, thereby curtailing each model's capacity to learn independently. This phenomenon, known as gradient contention, occurs when gradients produced by different models or tasks conflict with each other, leading to ineffective optimization. Specifically, when gradients push shared model parameters in opposing directions, they may neutralize each other's effect, hindering both models' learning processes. To preclude gradient contention across models and, at the same time, enable effective knowledge transmission and reinforce each model's autonomy in optimization, we devise an iterative training algorithm. This algorithm partitions loss optimization into several phases and carries them out in a sequentially alternating fashion. Initially, the procedure optimizes Model A's supervised loss L_{sup}^{A} while holding Model B's parameters static, thus allowing Model A to refine its performance under the supervision of true labels; subsequently, it optimizes Model B's supervised loss L_{sup}^{B} while freezing Model A's parameters. Thereafter, the framework employs the transfer losses L_{trans}^{A} and L_{trans}^{B} to impose knowledge constraints on Models A and B, respectively, using pseudo-labels to guide inter-model refinement. See Algorithm 1 for details. This staggered iterative update scheme safeguards each model's independent feature learning, using pseudo-labels to facilitate a more adaptable form of collaboration. In the end, after repeated iterations, the predictions from both models increasingly align, culminating in a synergistic improvement in overall performance. This algorithm deftly circumvents gradient contention while bolstering both models' learning capabilities and operational robustness.

Algorithm 1: Dynamic Loss Adjustment with IoU-based Weighting

Input: $[t_s^A, t_e^A]$, $[t_s^B, t_e^B]$: Predicted ranges for models A and B, GT: Ground Truth, λ: IoU threshold, K: Iterations
Output: Optimized parameters for models A and B
Initialization: Initialize $A, B, \eta_A, \eta_B, \alpha, \beta, \gamma$, and ω;
for $k \leftarrow 1$ ***to*** K **do**
  Compute IoU: $\mathrm{IoU}(A,B) \leftarrow \frac{\max\left(0, \min(t_e^A, t_e^B) - \max(t_s^A, t_s^B)\right)}{(t_e^A - t_s^A) + (t_e^B - t_s^B) - \max\left(0, \min(t_e^A, t_e^B) - \max(t_s^A, t_s^B)\right)}$;
  if $IoU(A,B) > \lambda$ **then**
    $\omega \leftarrow \gamma \cdot (1 - \mathrm{IoU}(A,B))$;
    $L_{\text{sup}}^A \leftarrow \mathcal{L}(P^A, GT)$;
    $A \leftarrow A - \eta_A \cdot \nabla L_{\text{sup}}^A$;
    $L_{\text{sup}}^B \leftarrow \mathcal{L}(P^B, GT)$;
    $B \leftarrow B - \eta_B \cdot \nabla L_{\text{sup}}^B$;
  else
    $\omega \leftarrow 1 - \mathrm{IoU}(A,B)$;
    $L_{\text{sup}}^A \leftarrow \mathcal{L}(P^A, GT)$, $L_{\text{trans}}^A \leftarrow \mathcal{L}(P^A, PL^B)$;
    $A \leftarrow A - \eta_A \cdot \nabla(L_{\text{sup}}^A + \alpha \cdot \omega \cdot L_{\text{trans}}^A)$;
    $L_{\text{sup}}^B \leftarrow \mathcal{L}(P^B, GT)$, $L_{\text{trans}}^B \leftarrow \mathcal{L}(P^B, PL^A)$;
    $B \leftarrow B - \eta_B \cdot \nabla(L_{\text{sup}}^B + \beta \cdot \omega \cdot L_{\text{trans}}^B)$;
  `// Compute total loss for joint optimization`
  $LOSS \leftarrow L_{\text{sup}}^A + L_{\text{sup}}^B + \alpha \cdot \omega \cdot L_{\text{trans}}^A + \beta \cdot \omega \cdot L_{\text{trans}}^B$;
  Jointly update A and B: $A, B \leftarrow$ Optimize using $LOSS$;
return Optimized parameters for A and B

4. Experiments

4.1. Datasets

To evaluate the performance of our proposed MMCL, we conduct experiments on three challenging video temporal grounding datasets:

Charades-STA [1] consists of videos portraying routine activities, sourced from the Charades dataset [37]. It encompasses 6672 videos, 16,128 annotations, and 11,767 temporal moments. The uneven distribution of clip lengths and a high degree of overlap place considerable demands on local feature modeling. Of the annotated moments, 12,408 are allocated for training and 3720 for testing.

ActivityNet Caption [38] was initially constructed for dense video captioning. It comprises around 20,000 YouTube videos with diverse contents, averaging 120 s in length. The dataset features semantic variety and broad domains, placing higher demands on the multimodal representation capability of models. Following the experimental setup in [22,23], val1 is used as the validation set and val2 as the test set, containing 37,421 and 17,505 annotated temporal moments for training and testing, respectively. Currently, this is the largest dataset in this task.

TACoS [39] is a video temporal grounding dataset focusing on kitchen scenarios, derived from the MPII Cooking Composite Activities dataset [39]. It comprises 127 videos with annotated time segments, totaling over 18 h in duration. The dataset features a large number of consecutive actions, with dense temporal distributions and a high overlap ratio, posing greater challenges for boundary-based models. Following [22,23], 10,146, 4589, and 4083 temporal annotations are used for training, validation, and testing, respectively.

4.2. Baselines

We employ VSLNet [22] and 2D-TAN [23] as two baseline networks in the framework, representing the typical Span-based and 2D-MAP models, respectively. Both have publicly available source code, and we train and evaluate the MMCL framework on the three datasets described in Section 4.1.

4.3. Implementation Details

4.3.1. Evaluation Metrics

Following existing video grounding works, we evaluate the performance on two main metrics:

mIoU: This metric measures the mean Intersection over Union (IoU) between the model's predicted intervals and the reference intervals. By averaging the IoU values over the entire test set, it reflects the overall localization accuracy of the model and effectively evaluates the precision of temporal segment predictions.

Recall: We adopt "$Rank@n, \text{IoU} = m$" as the evaluation metrics, following [1]. The "$Rank@n, \text{IoU} = m$" represents the percentage of language queries having at least one result whose IoU between $top - n$ predictions with ground-truth is larger than m. In our experiments, we reported the results of $n = 1$ and $m \in \{0.3, 0.5, 0.7\}$.

4.3.2. Experimental Settings

For each baseline method, VSLNet follows the configurations reported in their original paper, adopting the C3D-feature version of VSLNet. As for 2D-TAN, the components of the framework likewise follow the same parameters as presented in its original documentation. The hyperparameters α and β (refer to Section 3.4) regulate the weighting for the transfer losses L^A_{trans} and L^B_{trans}. Based on validation tuning, we choose $\alpha = 0.5$ and $\beta = 0.5$ to strike a balance between supervision and transfer objectives in model optimization. The IoU threshold λ (see Section 3.5) evaluates predictive alignment between models, set at 0.5 to maintain a reasonable level of discriminatory power. The factor γ (see Section 3.5), responsible for dynamically modulating weights, is fixed at 0.8 to attenuate the transfer loss under scenarios of high model consistency. The number of video frames n_f in the CLIP-Guided Module (see Section 3.3.1) is set to 32 to balance semantic representational power and computational efficiency. The similarity threshold τ (see Section 3.3.1) is set to 0.4, filtering out highly relevant frames to generate a more precise guiding temporal range. The learning rates η_A and η_A are both set to 0.0001, and the number of iterative optimization cycles K (see Algorithm 1) is set to 10.

4.4. Training and Inference

This subsection provides an examination of how the MMCL framework performs throughout both training and inference stages. Figure 3 juxtaposes the framework's outcomes in loss optimization, test accuracy, signal learning, and noise memorization.

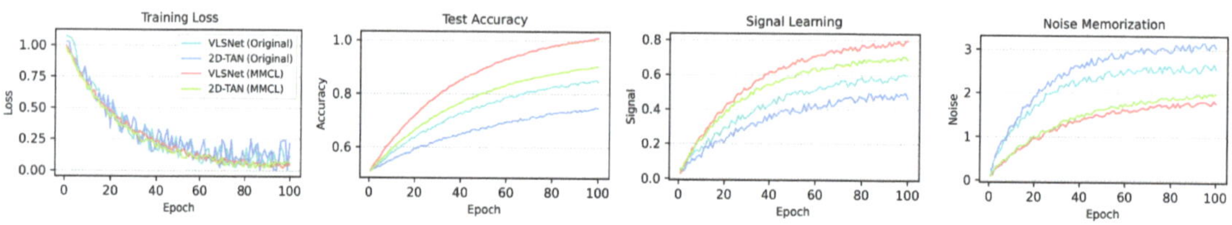

Figure 3. Comparison between single-model learning and multi-model collaborative learning in terms of training loss, test accuracy, signal learning, and noise memorization.

Observing the training loss curve reveals that MMCL demonstrates a noticeably swifter reduction in loss compared with the baseline model. In addition, as training epochs progress, the loss steadily converges to a more stable, lower plateau. This expedited convergence owes much to the framework's bidirectional knowledge transfer design, which allows the two models to exchange learned representations and leverage pseudo-labels as supplementary supervisory signals, thus steering clear of local optima. In the testing phase, the cooperative learning facilitated by the MMCL framework markedly enhances the accuracy of both models. Especially during the final training iterations, VLSNet's contextual reasoning and 2D-TAN's temporal boundary localization both become more refined, indicating that this collaborative mechanism can successfully broaden each model's performance ceiling. The signal learning trajectory reveals that the MMCL framework incrementally fortifies the cooperative models' capacity to capture relevant signals, subsequently stabilizing and outperforming any single-model baselines. CLIP's guidance mechanism, leveraging semantic priors to refine pseudo-labels' precision and coherence, plays a key role in these gains. The juxtaposition of noise memorization curves offers additional evidence of the efficacy of CLIP-guided pseudo-labels. By more effectively suppressing spurious or low-fidelity time segments, the MMCL framework supplies purer supervisory signals for training.

4.5. Comparison with State-of-the-Arts

4.5.1. Comparative Methods

We compare our proposed MMCL with state-of-the-art video temporal grounding methods on three public datasets. These methods can be grouped into two categories:

(1) Span-based Models: CTRL [1], ROLE [40], ACL [41], SM-RL [42], ExCL [43], DRN [44], CBLN [45], PS-VTG [28], D3G [46], PFU [47], TGN [48], ABLR-af [49], ABLR-aw [49], ACRN [50], VSLNet [22], MRTNet(VSLNet) [51];

(2) 2D-MAP Models: SAP [52], QSPN [53], DEBUG [54], MAN [55], GDP [56], CI-MHA [57], CMIN [58], 2D-TAN [23], MMN [59], MRTNet(2D-TAN) [51]

4.5.2. Quantitative Analysis

The outcomes on the three benchmarks, as reported in Tables 1–3, indicate that our MMCL framework significantly enhances the baseline models' performance.

Summarizing the results in Tables 1–3, we see that on the Charades-STA dataset, 2D-TAN underperforms VSLNet mainly because its 2D-MAP mechanism struggles with short segments and high-overlap video scenarios, whereas VSLNet's boundary-classification approach allows more precise modeling of short-segment boundaries. After incorporating the MMCL framework, 2D-TAN learns the semantic modeling capability of VSLNet's boundary-classification approach, improving its mIoU to 42.17 and achieving a relative improvement of 7.5%, demonstrating that the MMCL framework effectively compensates for 2D-TAN's shortcomings in modeling short segments. By contrast, on the TACoS dataset, VSLNet underperforms 2D-TAN because its span-based structure finds it difficult to handle the dense and uneven segment distribution of TACoS, while 2D-TAN's 2D-MAP mechanism is better suited for complex temporal-relationship modeling. After integrating the MMCL framework, VSLNet acquires long-segment relational modeling and temporal context-capturing abilities from 2D-TAN, raising its mIoU to 27.83 and yielding a 15.4% relative improvement. These results demonstrate that the MMCL framework, through knowledge transfer and collaborative learning, significantly improves both models' performance on the datasets where they originally fared poorly.

Table 1. Performance comparison with the state-of-the-art models on Charades-STA dataset.

Methods	Publication	Rank@1			mIoU
		IoU = 0.3	IoU = 0.5	IoU = 0.7	
CTRL [1]	ICCV2017	–	23.63	8.9	–
ROLE [40]	MM2018	25.26	12.12	–	–
ACL [41]	WACV2019	–	30.48	12.20	–
SAP [52]	AAAI2019	–	27.42	13.36	–
SM-RL [42]	CVPR2019	–	24.36	11.17	–
QSPN [53]	AAAI2019	54.70	35.60	15.80	–
DEBUG [54]	EMNLP2019	54.95	37.39	17.92	36.34
ExCL [43]	ACL2019	–	44.10	22.40	–
MAN [55]	CVPR2019	–	46.53	22.72	–
GDP [56]	AAAI2020	54.54	39.47	18.49	–
DRN [44]	CVPR2020	–	42.90	23.68	–
CBLN [45]	CVPR2021	–	43.67	24.44	–
CI-MHA [57]	SIGIR2021	69.87	54.68	35.27	–
MMN [59]	AAAI2022	–	47.31	27.28	–
PS-VTG [28]	ITM2022	–	39.22	20.17	–
D3G [46]	ICCV2023	–	41.64	19.60	–
2D-TAN [23]	AAAI2020	57.31	39.70	23.31	39.23
MRTNet (2D-TAN) [51]	ICASSP2024	59.23	44.27	25.88	40.59
MMCL (2D-TAN)	–	60.58	46.95	28.26	42.17
VSLNet [22]	ACL2020	70.46	54.19	35.22	50.02
MRTNet (VSLNet) [51]	ICASSP2024	70.88	56.19	36.37	50.74
MMCL (VSLNet)	–	**72.31**	**58.86**	**38.52**	**51.96**

Table 2. Performance comparison with the state-of-the art models on ActivityNet Caption dataset.

Methods	Publication	Rank@1			mIoU
		IoU = 0.3	IoU = 0.5	IoU = 0.7	
TGN [48]	EMNLP2018	43.81	27.93	–	–
CMIN [58]	SIGIR2019	63.61	43.40	23.88	–
QSPN [53]	AAAI2019	45.30	27.70	13.60	–
ABLR-af [49]	AAAI2019	53.65	34.91	–	35.72
ABLR-aw [49]	AAAI2019	55.67	36.79	–	36.99
DEBUG [54]	EMNLP2019	55.91	39.72	–	39.51
GDP [56]	AAAI2020	56.17	39.27	–	39.80
DRN [44]	CVPR2020	58.52	41.51	23.07	43.13
CI-MHA [57]	SIGIR2021	61.49	43.97	25.13	–
SeqPAN [60]	ACL2021	61.65	45.50	29.37	45.11
CBLN [45]	CVPR2021	66.34	48.12	27.60	–
MMN [59]	AAAI2022	65.05	48.59	29.26	–
PS-VTG [28]	ITM2022	59.71	39.59	21.98	–
PFU [47]	ARXIV2023	59.63	36.35	16.61	40.15
D3G [46]	ICCV2023	58.25	36.68	18.54	–
2D-TAN [23]	AAAI2020	59.45	44.51	26.54	43.29
MRTNet (2D-TAN) [51]	ICASSP2024	60.71	45.59	28.07	44.54
MMCL (2D-TAN)	–	62.15	47.21	29.35	46.53
VSLNet [22]	ACL2020	63.16	43.22	26.16	43.19
MRTNet (VSLNet) [51]	ICASSP2024	64.17	44.09	27.43	44.82
MMCL (VSLNet)	–	65.59	46.82	28.59	45.98

Table 3. Performance comparison with the state-of-the-art models on TACoS dataset.

Methods	Publication	Rank@1			mIoU
		IoU = 0.3	IoU = 0.5	IoU = 0.7	
CTRL [1]	ICCV2017	18.32	13.30	–	–
ACRN [50]	SIGIR2018	19.52	14.62	–	–
TGN [48]	EMNLP2018	21.77	18.90	–	–
SM-RL [42]	CVPR2019	20.25	15.95	–	–
ACL [41]	WACV2019	22.07	17.78	–	–
ABLR-aw [49]	AAAI2019	18.90	9.30	–	12.50
DEBUG [54]	EMNLP2019	23.45	11.72	–	16.03
CMIN [58]	SIGIR2019	24.64	18.05	–	–
CBLN [45]	CVPR2021	38.98	27.65	–	–
SeqPAN [60]	ACL2021	31.72	27.19	21.65	25.86
MMN [59]	AAAI2022	38.57	27.24	–	–
PS-VTG [28]	ITM2022	23.64	10.00	3.35	–
D3G [46]	ICCV2023	27.27	12.67	4.70	–
2D-TAN [23]	AAAI2020	37.29	25.32	13.32	25.19
MRTNet (2D-TAN) [51]	ICASSP2024	37.81	26.01	14.95	26.29
MMCL (2D-TAN)	–	**39.50**	**27.50**	15.12	**27.85**
VSLNet [22]	ACL2020	29.61	24.27	20.03	24.11
MRTNet (VSLNet) [51]	ICASSP 2024	32.35	25.84	21.31	26.14
MMCL (VSLNet)	–	34.15	26.81	22.57	27.83

From the results in Table 2, both models achieve improved performance on the ActivityNet Caption dataset. With its semantic modeling advantage, VSLNet achieves an mIoU of 43.19 on this dataset. However, by learning richer temporal relationship modeling from 2D-TAN through the MMCL framework, its mIoU rises to 45.98, a relative improvement of 6.4%. Similarly, by sharing knowledge with VLSNet, 2D-TAN advances its mIoU from 43.29 to 46.53, achieving a 7.1% relative improvement.

In sum, through cross-model knowledge sharing and joint learning, the MMCL framework allows the two baseline models to mutually address their shortcomings and amplify their respective advantages. On TACoS, VLSNet reaps the benefits of 2D-TAN's temporal relationship modeling; on Charades-STA, 2D-TAN inherits stronger boundary-modeling capabilities from VLSNet. This reciprocal learning and knowledge transfer characteristic endows the MMCL framework with robust adaptability across different task scenarios. Moreover, relative to other competing methods and MRTNet, the MMCL framework achieves superior results on all metrics in each of the three datasets, offering additional validation of its efficacy.

4.6. Performance of Isomorphic Models

Even under identical model architectures, the random sampling of training data and mini-batch dynamics induce divergent biases in temporal segments, feature representations, or modality alignment. Such discrepancies lay the groundwork for knowledge transfer between structurally identical models. To assess the MMCL framework's efficacy and potential constraints in same-architecture collaborative learning, we designate 2D-TAN as both Model A and Model B and conduct experiments on the Charades-STA dataset, employing the hyperparameters outlined in Section 4.3.2. According to Figure 4, the models utilizing the MMCL framework demonstrate a degree of performance enhancement relative to the baseline model. This finding suggests that even with a uniform model design, differences in data inputs and the stochastic nature of feature learning give rise to variations in predictive outcomes. The MMCL framework capitalizes on these variations for inter-model knowledge transfer, thereby achieving a synergistic boost in overall performance.

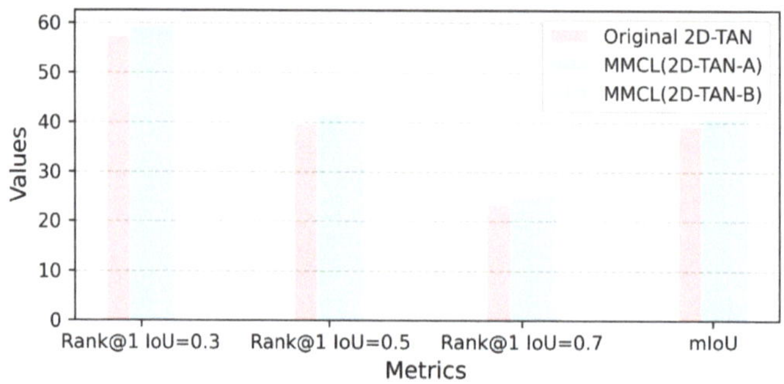

Figure 4. Performance comparison of isomorphic models on the Charades-STA dataset under the MMCL framework.

4.7. Ablation Studies

In an effort to streamline the experimental design, the following ablation studies are carried out on the Charades-STA dataset, leveraging VSLNet and 2D-TAN as baseline architectures.

4.7.1. Loss Function

To evaluate the effectiveness of each component in the proposed loss function, we conduct a series of tests with different loss settings, as shown in Table 4.

Table 4. Performance comparison of MMCL with different losses on Charades-STA dataset.

Experimental Setting		Rank@1			mIoU
		IoU = 0.3	IoU = 0.3	IoU = 0.3	
MMCL (VSLNet)	Ours	72.31	58.86	38.52	51.96
	w/o L_{sup}^{A}	68.56	52.82	33.89	48.11
	w/o L_{sup}^{B}	70.12	56.71	36.25	49.92
	w/o L_{trans}^{A}	70.85	56.71	36.79	50.25
	w/o L_{trans}^{B}	71.92	57.91	37.65	51.35
MMCL (2D-TAN)	Ours	60.58	46.95	28.26	42.17
	w/o L_{sup}^{A}	59.41	45.63	27.11	41.03
	w/o L_{sup}^{B}	55.81	38.89	22.18	37.84
	w/o L_{trans}^{A}	59.83	46.42	27.94	41.72
	w/o L_{trans}^{B}	58.78	45.56	27.21	40.85

Supervision loss (L_{sup}^{A} and L_{sup}^{B}) underpins the independent optimization of each model; removing this component forces the affected model to depend predominantly on transfer loss and pseudo-labels for training. As an example, omitting L_{sup}^{A} strips Model A of its capacity for standalone optimization, resulting in a notable performance drop. Concurrently, the pseudo-labels transferred to Model B deteriorate in quality, undermining the efficacy of knowledge transfer. However, since Model B still retains L_{sup}^{B}, its performance decline is relatively limited. Similarly, removing L_{sup}^{B} leads to a significant reduction in Model B's performance, especially when capturing strict boundaries (e.g., $IoU = 0.7$).

Model A is also affected by the degraded quality of pseudo-labels, but to a lesser extent, as it still depends on L_{sup}^{A} . Excluding the transfer losses (L_{trans}^{A} and L_{trans}^{B}) disrupts the inter-model knowledge-sharing mechanism. Removing L_{trans}^{A} prevents Model A from assimilating supplementary insights from Model B, leading to a performance dip. Model B, retaining L_{trans}^{B} and its corresponding supervisory signals, suffers only marginal setbacks. Conversely, removing L_{sup}^{B} robs Model B of its ability to acquire knowledge from Model A, causing a significant performance drop, whereas Model A's performance sees only a minor decline since it can still rely on its own supervision signals for independent optimization.

4.7.2. Hyperparameters of α and β

From Equation (22) in Section 3.5.2, it is evident that the base weights α and β for the transfer loss exert a direct influence on the strength of inter-model knowledge transfer. In this study, we hold γ constant while varying the fundamental weights α and β. Referring to Figure 5, once α and β approximate 0.5, knowledge transfer finds an equilibrium that enables mutual learning and leads to superior overall performance. When α and β remain small, the force of knowledge transfer lessens, causing each model to rely heavily on its own supervision and thereby producing results akin to the baseline model. This situation suggests that inter-model cooperation is curtailed, and it becomes challenging to compensate for the limitations inherent in a solitary model. If α and β are severely skewed (e.g., α is much larger than β or vice versa), knowledge transfer adopts a one-sided flow, resulting in asymmetrical performance. Conversely, if α and β are both sizable, the frequency of knowledge transfer escalates, injecting abundant "noise" that hampers the model's ability to distinguish useful supervisory cues from extraneous transfer signals, ultimately diminishing cooperative gains. Taken together, properly tuning α and β is essential for upholding equilibrium and moderation in inter-model knowledge sharing; doing so maximizes each model's unique advantages while averting skewed transfer or excessive disruption that might undermine overall performance.

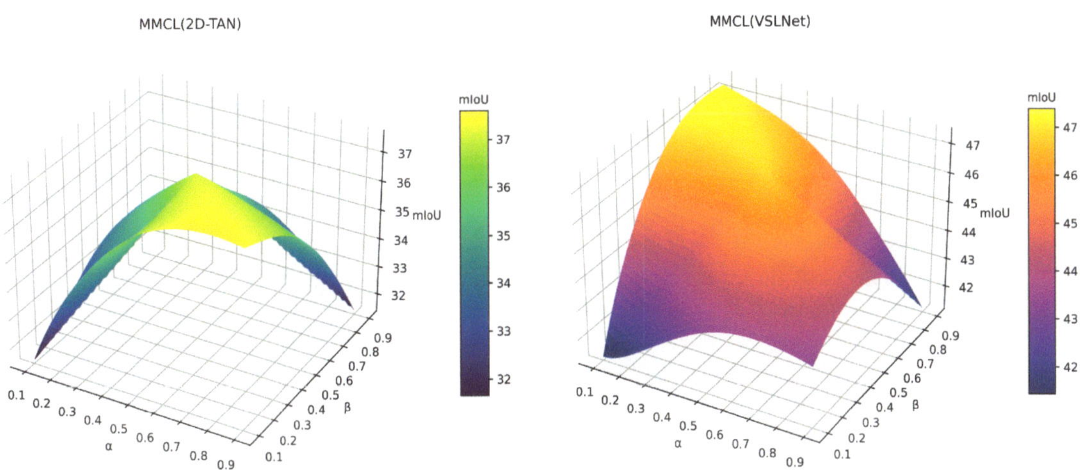

Figure 5. Effect of transfer-loss weights (α and β) on model performance.

4.7.3. Hyperparameters of λ and γ

The hyperparameters λ and γ, which govern the dynamic adjustment of transfer loss (see Section 3.5.2), play key roles in determining recall, precision, and overall inter-model cooperation. In this study, α and β remain fixed at 0.5, while performance metrics are gathered for various $\gamma - \lambda$ pairs (see Figure 6). As the IoU consistency threshold, λ measures the degree of similarity in the models' predictions. A relatively small λ defines consistency more leniently, capturing a greater number of predictions and thereby boosting recall, yet at the risk of admitting low-quality knowledge flow that can erode precision. By contrast, a larger λ imposes tighter consistency requirements, fostering higher precision but limiting collaborative interactions between models and thereby diminishing recall. Experimental findings suggest that a suitably moderate λ strikes an effective compromise between shared knowledge transfer and independent optimization. Meanwhile, the dynamic factor γ dictates the degree of transfer-loss attenuation when models exhibit high consistency. A larger γ enables models to lean more on their internal supervision signals, thus bolstering precision. Conversely, a smaller γ strengthens cross-model knowledge sharing, likely improving recall at the expense of reduced independence in optimization. Under lenient consistency requirements (e.g., *Rank*@1, *IoU* $= 0.3$), the influence of λ and γ on recall is minimal, producing relatively stable performance. This suggests that collaborative knowledge exchange reliably enhances the model's overall recall capability. Under more balanced consistency thresholds (e.g., *Rank*@1, *IoU* $= 0.5$), the models grow increasingly sensitive to variations in λ and γ, making prudent parameter choices pivotal for balancing recall and precision. Under stringent consistency metrics (e.g., *Rank*@1, *IoU* $= 0.7$), the models display the highest sensitivity to λ and γ, and moderate λ in combination with a mid-to-high γ yields peak performance during collaborative learning. In summary, λ and γ jointly manage the trade-off between recall and precision and, by dynamically calibrating the strength of knowledge transfer, serve as key levers for enhancing the overall efficacy of cooperative learning.

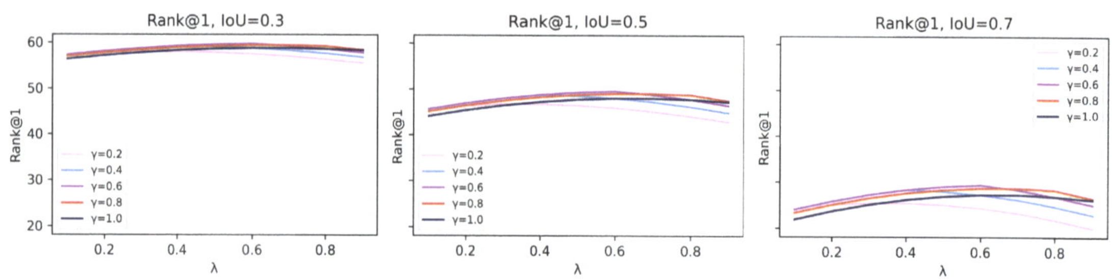

Figure 6. Effect of the consistency threshold (λ) and the dynamic adjustment factor (γ) on model performance.

4.7.4. Effectiveness Analysis of the CLIP-Guided Module

In the CLIP-guided module, both the number of sampled video frames n_f and the similarity threshold τ play a key role in determining the quality of pseudo-labels, thereby shaping the outcomes of inter-model knowledge transfer as well as final performance. In this subsection, we separately discuss the impact of varying the number of video frames and the similarity threshold on the MMCL framework, with the corresponding results presented in Figure 7. First, holding the similarity threshold constant at $\tau = 0.4$, we gradually raise the number of sampled frames. As depicted in Figure 7a, the model's performance (mIoU) progressively increases alongside the number of frames. This indicates that a higher frame count enriches the temporal scope and contextual fidelity of the pseudo-labels, effectively expanding the reservoir of knowledge they convey. Nevertheless, upon reaching 64 frames, performance gains level off or marginally drop because incorporating excessive frames does not translate into significant feature benefits but rather introduces unwarranted noise and squanders computational overhead. Subsequently, keeping $n_f = 32$ constant, we adjust the similarity threshold τ to gauge how performance responds. As illustrated in

Figure 7b, the model's performance follows a nonlinear trajectory in response to variations in the similarity threshold τ. A too-low threshold incorporates excessive low-grade segments in the pseudo-labels, weakening their utility in guiding optimization. Conversely, a too-stringent threshold narrows the scope of pseudo-labels, hindering the capture of the video's broader context and thereby stifling collaborative improvement. In comparison, a moderate threshold ($\tau = 0.4$) attains an equilibrium between filtering out noise and retaining adequate coverage, yielding pseudo-labels of superior quality for facilitating knowledge transfer. In sum, judiciously determining the number of frames and the similarity threshold is vital for producing robust pseudo-labels, thereby proving indispensable for successful inter-model knowledge transfer and performance enhancement.

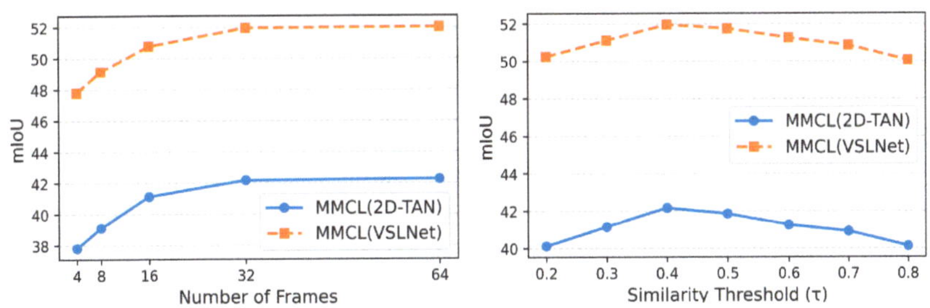

(**a**) Impact of number of frames on mIoU. (**b**) Impact of similarity threshold on mIoU.

Figure 7. Effect of the number of video frames and the similarity threshold on performance in the CLIP-guided module.

4.7.5. Qualitative Analysis

Figure 8 shows qualitative results of MMCL(VSLNet) on the Charades-STA dataset and MMCL(2D-TAN) on the TACoS dataset. From the examples, we can observe that in the Charades-STA dataset, which features short action intervals with rapidly shifting boundaries, MMCL(VSLNet) demonstrates heightened temporal sensitivity and refined boundary delineation under CLIP guidance. As for the TACoS dataset, known for its longer action durations, MMCL(2D-TAN) demonstrates greater boundary-capturing precision under CLIP guidance. In the absence of CLIP guidance, the model frequently yields more expansive time spans than the actual annotations, hindering its ability to precisely identify the critical moments aligned with the query. This result reaffirms the crucial function of the CLIP-guided module in refining pseudo-label accuracy.

4.8. Case Study

The case study evaluates the MMCL (Multi-Model Collaborative Learning) framework's ability to harness the complementary strengths of VSLNet and 2D-TAN models across various video temporal grounding scenarios, including short and long videos, complex temporal relationships, and fast-changing scenes.

In scenarios with short videos (Figure 9, left), where precise boundary detection is critical, MMCL improved 2D-TAN's performance by allowing it to learn from VSLNet's accuracy in temporal boundary prediction. Conversely, in long video scenarios (Figure 9, right), which require a broader temporal context, MMCL enhanced VSLNet's performance by incorporating 2D-TAN's strength in extended temporal modeling. This demonstrates the framework's capacity to facilitate effective mutual learning between models with distinct advantages. In Figure 10, the performance on "difficult" samples was examined through complex temporal scenarios and fast-changing scenes. Difficult samples in our study refer to scenarios that require a deep understanding of extended temporal dependencies or rapid visual changes. The task demands not only temporal segmentation but also an understanding of the sequential order of events. VSLNet, typically challenged by such scenarios, showed notable improvement under the MMCL framework, demonstrating enhanced temporal reasoning by leveraging 2D-TAN's strengths. Conversely, in fast-changing scenarios like the BMX race video, which required precise alignment between fast-paced visual cues and the query, MMCL (2D-TAN) showed a significant performance gain. This result indicates that 2D-TAN, through bidirectional knowledge transfer, learned from VSLNet's dynamic attention mechanism, which is particularly effective in fast-changing environments.

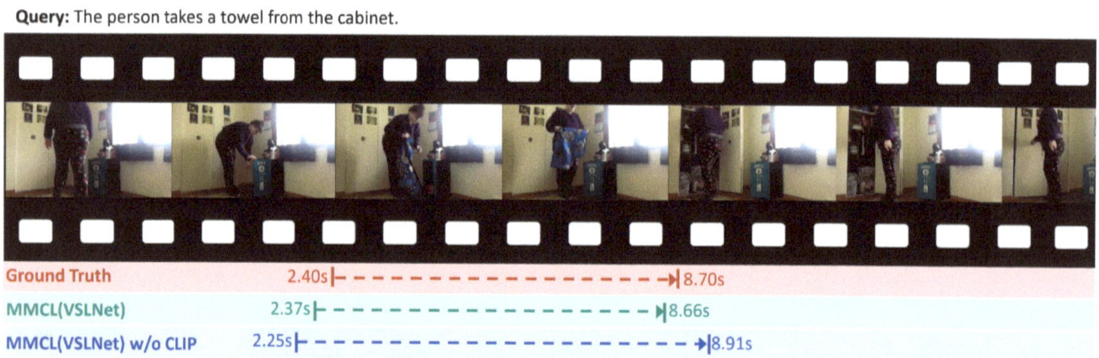

(**a**) Qualitative results on the Charades-STA dataset.

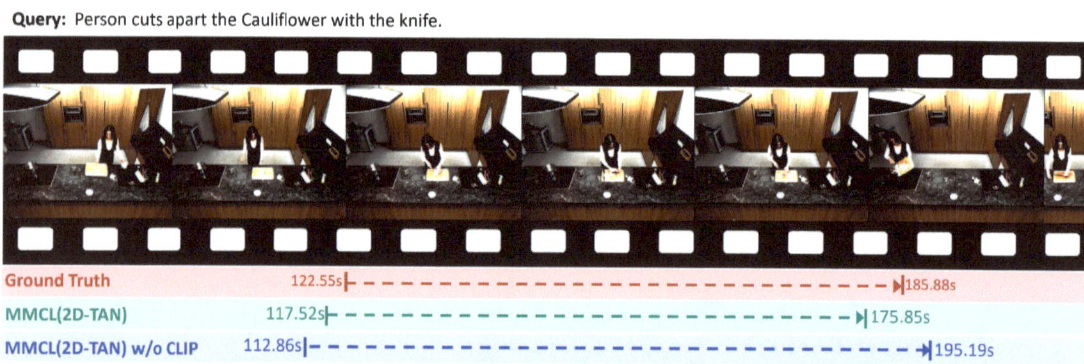

(**b**) Qualitative results on the TACoS dataset.

Figure 8. Qualitative results of MMCL(VSLNet) and MMCL(2D-TAN).

In specific challenging samples (Figures 11 and 12), the standalone models struggled with accurate predictions, whereas the MMCL-enhanced models succeeded. The MMCL (2D-TAN) model adapted better to fast-changing scenarios by learning from VSLNet's strengths, while MMCL (VSLNet) demonstrated improved performance in complex temporal scenarios by leveraging 2D-TAN's proficiency in modeling long-range temporal dependencies. These results confirm that the MMCL framework not only preserves each model's inherent strengths but also facilitates adaptive knowledge transfer, leading to improved performance on difficult samples and demonstrating the intended collaborative learning effect.

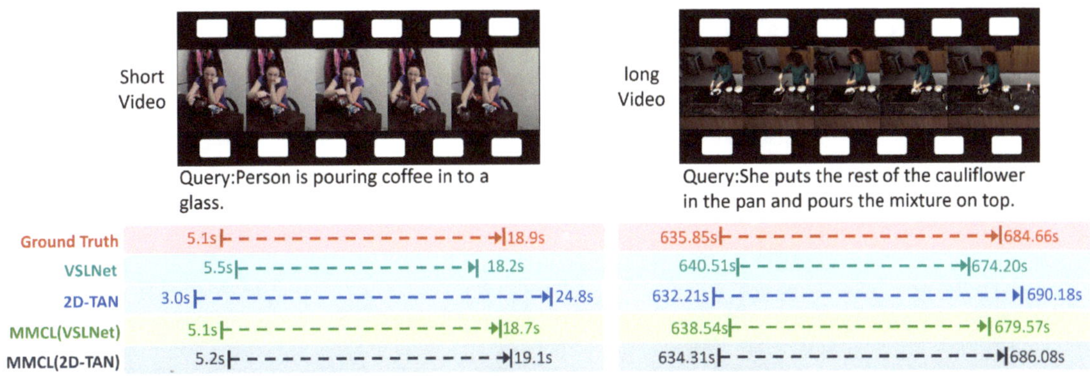

Figure 9. Performance of MMCL framework in short vs. long video scenarios.

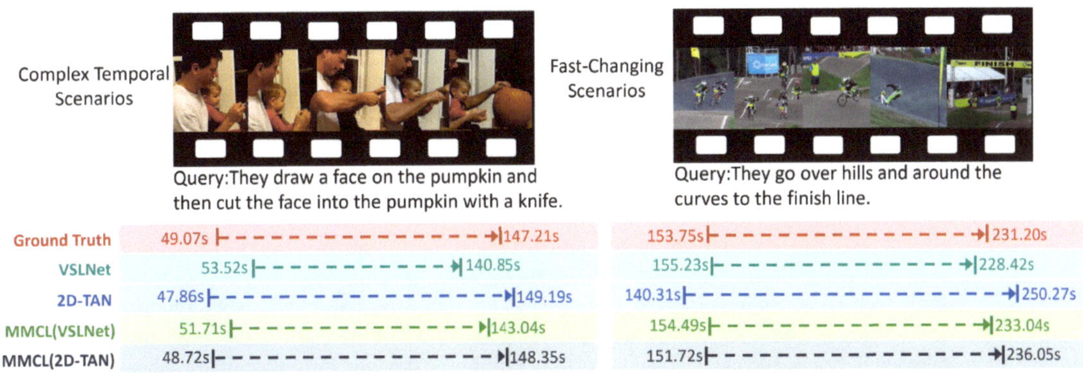

Figure 10. MMCL framework in fast-changing vs. complex temporal scenarios.

Figure 11. Improved temporal grounding in fast-changing scenarios with MMCL (2D-TAN).

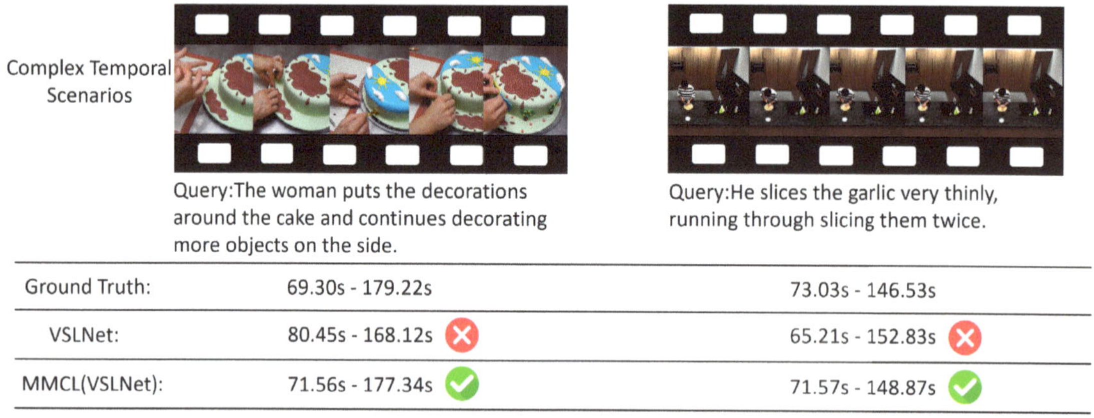

Figure 12. Enhanced performance in complex temporal scenarios with MMCL (VSLNet).

5. Conclusions

In this work, we proposed a Multi-Model Collaborative Learning (MMCL) framework that achieves efficient collaboration among multiple models through bidirectional knowledge transfer, a CLIP-guided pseudo-label generation mechanism, and an iterative training algorithm. Extensive evaluations across three benchmark datasets confirm that MMCL surpasses conventional baselines. Furthermore, the MMCL framework is highly versatile and can be employed in both isomorphic and heterogeneous model configurations, thus providing a novel paradigm for multi-model collaboration. While our current study focused on well-established datasets such as Charades-STA, ActivityNet Captions, and TACoS, we recognize the value of validating the MMCL framework on more recently released datasets, such as TVR and QVHighlight. These datasets introduce unique challenges, including longer video durations, more complex queries, and multi-view video sources. Exploring these datasets in future work will allow us to assess the generalization ability of MMCL under more diverse and evolving conditions. Even so, opportunities remain to enhance the framework's computational overhead and training efficiency. Future work will concentrate on devising a more compact architecture and real-time optimizations, broadening the applicability of the framework in diverse real-world contexts.

Author Contributions: Conceptualization, X.G. and J.W.; methodology, Y.T., X.G. and B.L.; software, Y.T. and B.L.; validation, Y.T. and B.L.; formal analysis, Y.T.; investigation, Y.T. and B.L.; resources, X.G. and J.W.; data curation, Y.T. and B.L.; writing—original draft preparation: Y.T. and B.L.; writing—review and editing: Y.T. and X.G.; visualization, Y.T.; supervision, X.G. and J.W.; project administration, X.G. and J.W.; funding acquisition, X.G. and S.Z. All authors have read and agreed to the published version of the manuscript.

Funding: This research was funded by the Natural Science Foundation of Guangdong Province (No. 2023A1515010673), in part by the Shenzhen Science and Technology Innovation Bureau key project (No. JSGG20220831110400001, No. CJGJZD20230724093303007, KJZD20240903101259001), in part by Shenzhen Medical Research Fund (No. D2404001), in part by Shenzhen Engineering Laboratory for Diagnosis & Treatment Key Technologies of Interventional Surgical Robots (XMHT20220104009), and the Key Laboratory of Biomedical Imaging Science and System, CAS, for the Research platform support.

Institutional Review Board Statement: Not applicable.

Informed Consent Statement: Not applicable.

Data Availability Statement: The data presented in this study are available on request from the corresponding author. The data are not publicly available due to restrictions in the data use agreement with the data provider.

Acknowledgments: The authors would like to thank all individuals and institutions who contributed to this research through technical support, administrative assistance, and valuable discussions. Their support is greatly appreciated.

Conflicts of Interest: There are no potential competing interests in our paper. All authors have seen the manuscript and approved to submit it to your journal. We confirm that the content of the manuscript has not been published or submitted for publication elsewhere.

References

1. Gao, J.; Sun, C.; Yang, Z.; Nevatia, R. TALL: Temporal Activity Localization via Language Query. In Proceedings of the IEEE International Conference on Computer Vision (ICCV), Venice, Italy, 22–29 October 2017.
2. Chu, Y.W.; Lin, K.Y.; Hsu, C.C.; Ku, L.W. End-to-End Recurrent Cross-Modality Attention for Video Dialogue. *IEEE/ACM Trans. Audio Speech Lang. Process.* **2021**, *29*, 2456–2464. [CrossRef]
3. Ji, W.; Li, Y.; Wei, M.; Shang, X.; Xiao, J.; Ren, T.; Chua, T.S. VidVRD 2021: The Third Grand Challenge on Video Relation Detection. In Proceedings of the 29th ACM International Conference on Multimedia, MM '21, Chengdu, China, 20–24 October 2021; pp. 4779–4783. [CrossRef]
4. Shang, X.; Li, Y.; Xiao, J.; Ji, W.; Chua, T.S. Video Visual Relation Detection via Iterative Inference. In Proceedings of the 29th ACM International Conference on Multimedia, MM '21, Chengdu, China, 20–24 October 2021; pp. 3654–3663. [CrossRef]
5. Shang, X.; Ren, T.; Guo, J.; Zhang, H.; Chua, T.S. Video Visual Relation Detection. In Proceedings of the 25th ACM International Conference on Multimedia, MM '17, Mountain View, CA, USA, 23–27 October 2017; pp. 1300–1308. [CrossRef]
6. Li, Y.; Wang, X.; Xiao, J.; Ji, W.; Chua, T.S. Invariant Grounding for Video Question Answering. In Proceedings of the IEEE/CVF Conference on Computer Vision and Pattern Recognition (CVPR), Los Alamitos, CA, USA, 19–24 June 2022; pp. 2928–2937.
7. Xiao, J.; Yao, A.; Liu, Z.; Li, Y.; Ji, W.; Chua, T.S. Video as Conditional Graph Hierarchy for Multi-Granular Question Answering. In Proceedings of the AAAI Conference on Artificial Intelligence, Online Conference, 22 February–1 March 2022; Volume 36, pp. 2804–2812. [CrossRef]
8. Li, S.; Li, B.; Sun, B.; Weng, Y. Towards Visual-Prompt Temporal Answer Grounding in Instructional Video. *IEEE Trans. Pattern Anal. Mach. Intell.* **2024**, *46*, 8836–8853. [CrossRef] [PubMed]
9. Liang, R.; Yang, Y.; Lu, H.; Li, L. Efficient Temporal Sentence Grounding in Videos with Multi-Teacher Knowledge Distillation. *arXiv* **2014**, arXiv:2308.03725.
10. Lan, X.; Yuan, Y.; Wang, X.; Wang, Z.; Zhu, W. A Survey on Temporal Sentence Grounding in Videos. *ACM Trans. Multimed. Comput. Commun. Appl.* **2023**, *19*, 1–33. [CrossRef]
11. Zhang, Y.; Xu, Y.; Chen, M.; Zhang, Y.; Feng, R.; Gao, S. SPTNET: Span-based Prompt Tuning for Video Grounding. In Proceedings of the 2023 IEEE International Conference on Multimedia and Expo (ICME), Brisbane, Australia, 10–14 July 2023; pp. 2807–2812. [CrossRef]
12. Ding, X.; Wang, N.; Zhang, S.; Huang, Z.; Li, X.; Tang, M.; Liu, T.; Gao, X. Exploring Language Hierarchy for Video Grounding. *IEEE Trans. Image Process.* **2022**, *31*, 4693–4706. [CrossRef] [PubMed]
13. Li, J.; Li, K.; Li, J.; Chen, G.; Wang, M.; Guo, D. Dual-path temporal map optimization for make-up temporal video grounding. *Multimed. Syst.* **2024**, *30*, 140. [CrossRef]
14. Zhang, Y.; Chen, X.; Jia, J.; Liu, S.; Ding, K. Text-Visual Prompting for Efficient 2D Temporal Video Grounding. In Proceedings of the IEEE/CVF Conference on Computer Vision and Pattern Recognition (CVPR), Vancouver, BC, Canada, 18–22 June 2023; pp. 14794–14804.
15. Bao, P.; Shao, Z.; Yang, W.; Ng, B.P.; Er, M.H.; Kot, A.C. Omnipotent Distillation with LLMs for Weakly-Supervised Natural Language Video Localization: When Divergence Meets Consistency. In Proceedings of the AAAI Conference on Artificial Intelligence, Vancouver, BC, Canada, 20–27 February 2024; Volume 38, pp. 747–755. [CrossRef]
16. Bao, P.; Xia, Y.; Yang, W.; Ng, B.P.; Er, M.H.; Kot, A.C. Local-Global Multi-Modal Distillation for Weakly-Supervised Temporal Video Grounding. In Proceedings of the AAAI Conference on Artificial Intelligence, Vancouver, BC, Canada, 20–27 February 2024; Volume 38, pp. 738–746. [CrossRef]
17. Weng, Y.; Li, B. Visual Answer Localization with Cross-Modal Mutual Knowledge Transfer. In Proceedings of the ICASSP 2023—2023 IEEE International Conference on Acoustics, Speech and Signal Processing (ICASSP), Rhodes Island, Greece, 4–10 June 2023; pp. 1–5. [CrossRef]

18. Wu, J.; Jiang, Y.; Wei, X.Y.; Li, Q. PolySmart @ TRECVid 2024 Medical Video Question Answering. *arXiv* **2024**, arXiv:2412.15514.
19. Shi, G.; Li, Q.; Zhang, W.; Chen, J.; Wu, X.M. Recon: Reducing Conflicting Gradients from the Root for Multi-Task Learning. *arXiv* **2023**, arXiv:2302.11289.
20. Liu, B.; Liu, X.; Jin, X.; Stone, P.; Liu, Q. Conflict-Averse Gradient Descent for Multi-task learning. In Proceedings of the Advances in Neural Information Processing Systems, Red Hook, NY, USA, 6–14 December 2021; Ranzato, M., Beygelzimer, A., Dauphin, Y., Liang, P., Vaughan, J.W., Eds.; Curran Associates, Inc.: Red Hook, NY, USA, 2021; Volume 34, pp. 18878–18890.
21. Radford, A.; Kim, J.W.; Hallacy, C.; Ramesh, A.; Goh, G.; Agarwal, S.; Sastry, G.; Askell, A.; Mishkin, P.; Clark, J.; et al. Learning Transferable Visual Models From Natural Language Supervision. In Proceedings of the 38th International Conference on Machine Learning, Virtual Event, 18–24 July 2021; Meila, M., Zhang, T., Eds.; Volume 139, pp. 8748–8763.
22. Zhang, H.; Sun, A.; Jing, W.; Zhou, J.T. Span-based Localizing Network for Natural Language Video Localization. *arXiv* **2020**, arXiv:2004.13931.
23. Zhang, S.; Peng, H.; Fu, J.; Luo, J. Learning 2D Temporal Adjacent Networks for Moment Localization with Natural Language. In Proceedings of the AAAI Conference on Artificial Intelligence, New York, NY, USA, 7–12 February 2020; Volume 34, pp. 12870–12877. [CrossRef]
24. Zhang, S.; Peng, H.; Fu, J.; Lu, Y.; Luo, J. Multi-Scale 2D Temporal Adjacency Networks for Moment Localization With Natural Language. *IEEE Trans. Pattern Anal. Mach. Intell.* **2022**, *44*, 9073–9087. [CrossRef] [PubMed]
25. Gou, J.; Yu, B.; Maybank, S.J.; Tao, D. Knowledge distillation: A survey. *Int. J. Comput. Vis.* **2021**, *129*, 1789–1819. [CrossRef]
26. Hinton, G. Distilling the Knowledge in a Neural Network. *arXiv* **2015**, arXiv:1503.02531.
27. Zheng, M.; Gong, S.; Jin, H.; Peng, Y.; Liu, Y. Generating Structured Pseudo Labels for Noise-resistant Zero-shot Video Sentence Localization. In Proceedings of the 61st Annual Meeting of the Association for Computational Linguistics (Volume 1: Long Papers), Toronto, ON, Canada, 9–14 July 2023; Rogers, A., Boyd-Graber, J., Okazaki, N., Eds.; pp. 14197–14209. [CrossRef]
28. Xu, Z.; Wei, K.; Yang, X.; Deng, C. Point-Supervised Video Temporal Grounding. *IEEE Trans. Multimed.* **2023**, *25*, 6121–6131. [CrossRef]
29. Xu, Y.; Wei, F.; Sun, X.; Yang, C.; Shen, Y.; Dai, B.; Zhou, B.; Lin, S. Cross-Model Pseudo-Labeling for Semi-Supervised Action Recognition. In Proceedings of the IEEE/CVF Conference on Computer Vision and Pattern Recognition (CVPR), New Orleans, Louisiana, 21–24 June 2022; pp. 2959–2968.
30. Jiang, X.; Xu, X.; Zhang, J.; Shen, F.; Cao, Z.; Shen, H.T. Semi-Supervised Video Paragraph Grounding With Contrastive Encoder. In Proceedings of the IEEE/CVF Conference on Computer Vision and Pattern Recognition (CVPR), New Orleans, Louisiana, 21–24 June 2022; pp. 2466–2475.
31. Luo, F.; Chen, S.; Chen, J.; Wu, Z.; Jiang, Y.G. Self-Supervised Learning for Semi-Supervised Temporal Language Grounding. *IEEE Trans. Multimed.* **2023**, *25*, 7747–7757. [CrossRef]
32. Piao, Y.; Lu, C.; Zhang, M.; Lu, H. Semi-Supervised Video Salient Object Detection Based on Uncertainty-Guided Pseudo Labels. In Proceedings of the Advances in Neural Information Processing Systems, New Orleans, LA, USA, 28 November–9 December 2022; Koyejo, S., Mohamed, S., Agarwal, A., Belgrave, D., Cho, K., Oh, A., Eds.; Curran Associates, Inc.: Red Hook, NY, USA, 2022; Volume 35, pp. 5614–5627.
33. Tran, D.; Bourdev, L.; Fergus, R.; Torresani, L.; Paluri, M. Learning Spatiotemporal Features With 3D Convolutional Networks. In Proceedings of the IEEE International Conference on Computer Vision (ICCV), Santiago, Chile, 7–13 December 2015.
34. Krizhevsky, A.; Sutskever, I.; Hinton, G.E. ImageNet Classification with Deep Convolutional Neural Networks. In Proceedings of the Advances in Neural Information Processing Systems 25 (NIPS 2012), Lake Tahoe, NV, USA, 3–6 December 2012; Pereira, F., Burges, C., Bottou, L., Weinberger, K., Eds.; Curran Associates, Inc.: Red Hook, NY, USA, 2012; Volume 25.
35. Pennington, J.; Socher, R.; Manning, C.D. Glove: Global vectors for word representation. In Proceedings of the 2014 Conference on Empirical Methods in Natural Language Processing (EMNLP), Doha, Qatar, 25–29 October 2014; pp. 1532–1543.
36. Alaparthi, S.; Mishra, M. Bidirectional Encoder Representations from Transformers (BERT): A sentiment analysis odyssey. *arXiv* **2020**, arXiv:2007.01127.
37. Sigurdsson, G.A.; Varol, G.; Wang, X.; Farhadi, A.; Laptev, I.; Gupta, A. Hollywood in homes: Crowdsourcing data collection for activity understanding. In Proceedings of the Computer Vision–ECCV 2016: 14th European Conference, Amsterdam, The Netherlands, 11–14 October 2016; Proceedings, Part I 14; Springer: Berlin/Heidelberg, Germany, 2016; pp. 510–526.
38. Caba Heilbron, F.; Escorcia, V.; Ghanem, B.; Carlos Niebles, J. ActivityNet: A Large-Scale Video Benchmark for Human Activity Understanding. In Proceedings of the IEEE Conference on Computer Vision and Pattern Recognition (CVPR), Boston, MA, USA, 7–12 June 2015.
39. Regneri, M.; Rohrbach, M.; Wetzel, D.; Thater, S.; Schiele, B.; Pinkal, M. Grounding Action Descriptions in Videos. *Trans. Assoc. Comput. Linguist.* **2013**, *1*, 25–36. [CrossRef]

40. Liu, M.; Wang, X.; Nie, L.; Tian, Q.; Chen, B.; Chua, T.S. Cross-modal Moment Localization in Videos. In Proceedings of the 26th ACM International Conference on Multimedia, MM '18, Seoul, Republic of Korea, 22–26 October 2018; pp. 843–851. [CrossRef]
41. Ge, R.; Gao, J.; Chen, K.; Nevatia, R. MAC: Mining Activity Concepts for Language-Based Temporal Localization. In Proceedings of the 2019 IEEE Winter Conference on Applications of Computer Vision (WACV), Waikoloa Village, HI, USA, 7–11 January 2019; pp. 245–253. [CrossRef]
42. Wang, W.; Huang, Y.; Wang, L. Language-Driven Temporal Activity Localization: A Semantic Matching Reinforcement Learning Model. In Proceedings of the IEEE/CVF Conference on Computer Vision and Pattern Recognition (CVPR), Long Beach, CA, USA, 15–20 June 2019.
43. Ghosh, S.; Agarwal, A.; Parekh, Z.; Hauptmann, A. ExCL: Extractive Clip Localization Using Natural Language Descriptions. *arXiv* **2019**, arXiv:1904.02755.
44. Zeng, R.; Xu, H.; Huang, W.; Chen, P.; Tan, M.; Gan, C. Dense Regression Network for Video Grounding. In Proceedings of the IEEE/CVF Conference on Computer Vision and Pattern Recognition (CVPR), Seattle, WA, USA, 13–19 June 2020.
45. Liu, D.; Qu, X.; Dong, J.; Zhou, P.; Cheng, Y.; Wei, W.; Xu, Z.; Xie, Y. Context-Aware Biaffine Localizing Network for Temporal Sentence Grounding. In Proceedings of the IEEE/CVF Conference on Computer Vision and Pattern Recognition (CVPR), Nashville, TN, USA, 20–25 June 2021; pp. 11235–11244.
46. Li, H.; Shu, X.; He, S.; Qiao, R.; Wen, W.; Guo, T.; Gan, B.; Sun, X. D3G: Exploring Gaussian Prior for Temporal Sentence Grounding with Glance Annotation. In Proceedings of the IEEE/CVF International Conference on Computer Vision (ICCV), Paris, France, 1–6 October 2023; pp. 13734–13746.
47. Ju, C.; Wang, H.; Liu, J.; Ma, C.; Zhang, Y.; Zhao, P.; Chang, J.; Tian, Q. Constraint and Union for Partially-Supervised Temporal Sentence Grounding. *arXiv* **2023**, arXiv:2302.09850.
48. Chen, J.; Chen, X.; Ma, L.; Jie, Z.; Chua, T.S. Temporally Grounding Natural Sentence in Video. In Proceedings of the 2018 Conference on Empirical Methods in Natural Language Processing, Brussels, Belgium, 31 October–4 November 2018; Riloff, E., Chiang, D., Hockenmaier, J., Tsujii, J., Eds.; pp. 162–171. [CrossRef]
49. Yuan, Y.; Mei, T.; Zhu, W. To Find Where You Talk: Temporal Sentence Localization in Video with Attention Based Location Regression. In Proceedings of the 2019 AAAI Conference on Artificial Intelligence, Hilton Hawaiian Village, Honolulu, HI, USA, 27 January–1 February 2019; pp. 9159–9166. [CrossRef]
50. Liu, M.; Wang, X.; Nie, L.; He, X.; Chen, B.; Chua, T.S. Attentive Moment Retrieval in Videos. In Proceedings of the 41st International ACM SIGIR Conference on Research & Development in Information Retrieval, SIGIR '18, Ann Arbor, MI, USA, 8–12 July 2018; pp. 15–24. [CrossRef]
51. Ji, W.; Qin, Y.; Chen, L.; Wei, Y.; Wu, Y.; Zimmermann, R. Mrtnet: Multi-Resolution Temporal Network for Video Sentence Grounding. In Proceedings of the ICASSP 2024—2024 IEEE International Conference on Acoustics, Speech and Signal Processing (ICASSP), Seoul, Republic of Korea, 14–19 April 2024; pp. 2770–2774. [CrossRef]
52. Chen, S.; Jiang, Y.G. Semantic Proposal for Activity Localization in Videos via Sentence Query. In Proceedings of the AAAI Conference on Artificial Intelligence, Hilton Hawaiian Village, Honolulu, HI, USA, 27 January–1 February 2019; pp. 8199–8206. [CrossRef]
53. Xu, H.; He, K.; Plummer, B.A.; Sigal, L.; Sclaroff, S.; Saenko, K. Multilevel Language and Vision Integration for Text-to-Clip Retrieval. In Proceedings of the AAAI Conference on Artificial Intelligence, Honolulu, HI, USA, 27 January–1 February 2019; Volume 33, pp. 9062–9069. [CrossRef]
54. Lu, C.; Chen, L.; Tan, C.; Li, X.; Xiao, J. DEBUG: A Dense Bottom-Up Grounding Approach for Natural Language Video Localization. In Proceedings of the 2019 Conference on Empirical Methods in Natural Language Processing and the 9th International Joint Conference on Natural Language Processing (EMNLP-IJCNLP), Hong Kong, China, November 2019; Inui, K., Jiang, J., Ng, V., Wan, X., Eds.; pp. 5144–5153. [CrossRef]
55. Zhang, D.; Dai, X.; Wang, X.; Wang, Y.F.; Davis, L.S. MAN: Moment Alignment Network for Natural Language Moment Retrieval via Iterative Graph Adjustment. In Proceedings of the IEEE/CVF Conference on Computer Vision and Pattern Recognition (CVPR), Long Beach, CA, USA, 15–20 June 2019.
56. Chen, L.; Lu, C.; Tang, S.; Xiao, J.; Zhang, D.; Tan, C.; Li, X. Rethinking the Bottom-Up Framework for Query-Based Video Localization. In Proceedings of the AAAI Conference on Artificial Intelligence, New York, NY, USA, 7–12 February 2020; pp. 10551–10558. [CrossRef]
57. Yu, X.; Malmir, M.; He, X.; Chen, J.; Wang, T.; Wu, Y.; Liu, Y.; Liu, Y. Cross Interaction Network for Natural Language Guided Video Moment Retrieval. In Proceedings of the 44th International ACM SIGIR Conference on Research and Development in Information Retrieval, SIGIR '21, Virtual, 11–15 July 2021; pp. 1860–1864. [CrossRef]
58. Zhang, Z.; Lin, Z.; Zhao, Z.; Xiao, Z. Cross-Modal Interaction Networks for Query-Based Moment Retrieval in Videos. In Proceedings of the 42nd International ACM SIGIR Conference on Research and Development in Information Retrieval, SIGIR'19, Paris, France, 21–25 July 2019; pp. 655–664. [CrossRef]

59. Wang, Z.; Wang, L.; Wu, T.; Li, T.; Wu, G. Negative Sample Matters: A Renaissance of Metric Learning for Temporal Grounding. In Proceedings of the AAAI Conference on Artificial Intelligence, Virtually, 22 February–1 March 2022; pp. 2613-2623. [CrossRef]
60. Zhang, H.; Sun, A.; Jing, W.; Zhen, L.; Zhou, J.T.; Goh, S.M.R. Parallel Attention Network with Sequence Matching for Video Grounding. In Proceedings of the Findings of the Association for Computational Linguistics: ACL-IJCNLP 2021, Online Event, 1–6 August 2021; Association for Computational Linguistics: Stroudsburg, PA, USA, 2021. [CrossRef]

applied sciences

MDPI

Article

Fine-Tuning Large Language Models for Ontology Engineering: A Comparative Analysis of GPT-4 and Mistral

Dimitrios Doumanas [1,*], Andreas Soularidis [1], Dimitris Spiliotopoulos [2], Costas Vassilakis [3] and Konstantinos Kotis [1,*]

1 Intelligent Systems Lab, Department of Cultural Technology and Communication, University of the Aegean, 81100 Mytilene, Greece; soularidis@aegean.gr
2 Department of Management Science and Technology, University of the Peloponnese, 22100 Tripolis, Greece; dspiliot@uop.gr
3 Department of Informatics and Telecommunications, University of the Peloponnese, 22100 Tripolis, Greece; costas@uop.gr
* Correspondence: cti23009@ct.aegean.gr (D.D.); kotis@aegean.gr (K.K.)

Abstract: Ontology engineering (OE) plays a critical role in modeling and managing structured knowledge across various domains. This study examines the performance of fine-tuned large language models (LLMs), specifically GPT-4 and Mistral 7B, in efficiently automating OE tasks. Foundational OE textbooks are used as the basis for dataset creation and for feeding the LLMs. The methodology involved segmenting texts into manageable chapters, generating question–answer pairs, and translating visual elements into description logic to curate fine-tuned datasets in JSONL format. This research aims to enhance the models' abilities to generate domain-specific ontologies, with hypotheses asserting that fine-tuned LLMs would outperform base models, and that domain-specific datasets would significantly improve their performance. Comparative experiments revealed that GPT-4 demonstrated superior accuracy and adherence to ontology syntax, albeit with higher computational costs. Conversely, Mistral 7B excelled in speed and cost efficiency but struggled with domain-specific tasks, often generating outputs that lacked syntactical precision and relevance. The presented results highlight the necessity of integrating domain-specific datasets to improve contextual understanding and practical utility in specialized applications, such as Search and Rescue (SAR) missions in wildfire incidents. Both models, despite their limitations, exhibited potential in understanding OE principles. However, their performance underscored the importance of aligning training data with domain-specific knowledge to emulate human expertise effectively. This study, based on and extending our previous work on the topic, concludes that fine-tuned LLMs with targeted datasets enhance their utility in OE, offering insights into improving future models for domain-specific applications. The findings advocate further exploration of hybrid solutions to balance accuracy and efficiency.

Keywords: large language models (LLMs) fine-tuning; ontology engineering (OE); domain-specific knowledge; search and rescue (SAR)

1. Introduction

OE is fundamental to the organization and structuring of semantic knowledge, which is increasingly essential in domains such as artificial intelligence, knowledge management, and digital information systems. Ontologies, by defining concepts and relations between them, serve as the backbone for systems that rely on contextual understanding and reasoning. As the volume and complexity of information grows, traditional manual approaches to OE are proving insufficient. This has led to the exploration of advanced methodologies for OE (especially concerning ontology learning) by leveraging machine learning methods and natural language processing (NLP).

LLMs, such as GPT-4 and Mistral, represent a significant leap in NLP capabilities, enabling automation in tasks that previously required extensive human expertise. By processing vast datasets, LLMs can perform nuanced tasks such as text summarization, content generation, and knowledge extraction. However, their application in OE—a domain requiring deep understanding of structure, logic, and domain-specific knowledge—remains underexplored. Despite their general utility, pre-trained LLMs often fail to meet the precision and relevance required for tasks such as ontology generation, primarily due to a lack of domain-specific training.

Fine-tuning involves a detailed procedure where the model's parameters are adjusted using a smaller, highly specialized dataset after the initial broad pre-training on extensive data. This targeted training approach helps in refining the model's predictions, making it adept at handling specific types of queries and generating outputs that adhere to the specialized knowledge structures typical in OE.

The effectiveness of fine-tuning is evident in the model's enhanced ability to navigate the intricacies of domain-specific languages and terminologies, enabling it to produce outputs that are not only grammatically correct but also contextually aligned with the domain's requirements. By leveraging fine-tuned LLMs, professionals in OE can automate the generation of ontological structures, significantly reducing the time and effort involved in manual ontology construction and potentially increasing the scalability of knowledge management practices across various domains.

Fine-tuning LLMs for OE tasks represents a confluence of advanced AI capabilities with the specialized needs of knowledge management. This process underpins significant improvements in how models understand and interact with complex, domain-specific datasets, marking a pivotal step towards more intelligent and automated systems in ontology management.

The primary challenge addressed in this research lies in the limitations of pre-trained LLMs when applied to OE, creating a gap in their ability to generate outputs with low F1 scores, reflecting poor precision and recall in representing domain-specific concepts and properties (both object and data properties). This shortcoming underscores the importance of fine-tuning LLMs to enhance their understanding of OE principles and methodologies.

The objective of this study is to bridge this gap by fine-tuning LLMs on carefully curated datasets derived from foundational texts in OE. These datasets, encompassing question–answer pairs, description logic, and translations of visual representations, are designed to align with the conceptual and structural needs of OE. The study evaluates the impact of fine-tuning on the models' ability to generate high-quality, domain-specific ontologies. This research is driven by two key research questions:

1. Can fine-tuning LLMs for OE concepts significantly improve their performance, enabling them to generate outputs with higher accuracy and adherence to ontology syntax?
2. Is the incorporation of domain-specific datasets into the fine-tuning process able to enhance the practical utility of the generated ontologies, particularly in real-world applications?

To answer these questions, comparative experiments were conducted on GPT-4 and Mistral 7b, two leading LLMs with distinct capabilities. These experiments assessed the models' performance in generating ontologies, focusing on metrics such as precision, recall, and F1 score. The results provide valuable insights into the capabilities and limitations of LLMs in OE and offer practical recommendations for enhancing their utility in this specialized field.

In this study, we opted to apply fine-tuning directly to the LLMs without first employing retrieval-augmented generation (RAG) to address domain-specific OE tasks. This decision was driven by the need for precision and tailored adaptation to the highly specialized domain requirements. Fine-tuning integrates the domain's specific characteristics into the model's internal parameters, which enhances its ability to produce contextually precise and terminologically consistent outputs. This is particularly critical for tasks such as OE, where maintaining strict adherence to predefined logical structures and workflows is paramount. Conversely, while RAG excels at dynamically retrieving and incorporating external knowledge, it introduces additional complexity and potential latency. Moreover, the reliance of RAG on external, often heterogeneous, data sources could lead to inconsistencies, inaccuracies, or even irrelevance in outputs, especially if the external knowledge is not meticulously curated. By prioritizing fine-tuning, we ensured greater control over the model's behavior and outputs, optimizing for performance in a domain characterized by stable knowledge frameworks rather than dynamic informational updates. This approach aligns with established evidence that fine-tuning is more effective than retrieval-based methods for domains requiring precision, efficiency, and consistent adherence to structured frameworks, making it the logical choice for our objectives in OE.

In addition to contributing to the understanding of LLMs' application in OE, this study highlights the broader implications of integrating machine learning with NLP for domain-specific tasks. By addressing the challenges of domain knowledge representation, this research sets the stage for future advancements in automating complex knowledge-driven processes, offering a pathway towards more intelligent, context-aware systems in diverse fields.

Overall, the goals of the presented work are as follows:

- Enhance OE with Fine-Tuned LLMs: To demonstrate how fine-tuning LLMs can improve their performance in generating and understanding ontologies aligned with formal principles and domain-specific requirements.
- Develop Domain-Specific OE Training Datasets: To create and utilize a structured methodology for curating datasets based on foundational OE-related handbooks, including question–answer pairs and translated visual elements, ensuring alignment with OE standards.
- Provide a Comparative Analysis of LLMs: To evaluate and contrast the performance of GPT-4 and Mistral 7B in terms of accuracy, efficiency, cost, and their ability to handle domain-specific OE tasks, providing insights into their practical applicability.
- Apply Fine-Tuned Models to Real-World Scenarios: To explore the use of fine-tuned LLMs in practical applications, such as Search and Rescue (SAR) missions during wildfires, showcasing the potential utility of these models in critical, specialized fields.
- Identifying Limitations and Future Directions: To critically assess the strengths and weaknesses of the models, emphasizing the need for hybrid solutions and domain-specific training data to enhance their contextual understanding and emulate human expertise effectively.
- Pave the Way for AI-Enhanced OE Tools: To advocate advancements in OE by integrating LLMs with domain-specific insights, aiming to bridge the gap between theoretical principles and practical utility. The approach demonstrates how fine-tuned LLMs can automate key ontology development tasks, including concept extraction, relationship identification, and hierarchical structuring. By leveraging pre-trained models, we reduce the reliance on manual curation, allowing for a more scalable and efficient OE process. This advancement is particularly relevant for domains requiring management systems, where rapid adaptation to evolving technologies is crucial.

By addressing these goals, this paper aims to advance the field of OE by leveraging fine-tuned LLMs, like GPT-4 and Mistral 7B, to enhance the automation and accuracy of ontology generation. It seeks to bridge the gap between theoretical principles and real-world applications by developing domain-specific datasets and applying the models to practical scenarios, such as SAR missions during wildfires. Through a detailed comparative analysis, the study provides actionable insights, offering guidance for selecting and optimizing models for specialized tasks. By highlighting the importance of domain-specific training and exploring hybrid solutions, the paper sets a benchmark for integrating LLMs into complex, knowledge-driven domains, ultimately aiming to inspire further innovation and provide practical tools for researchers and practitioners.

While this study focuses on fine-tuning large language models (LLMs) for ontology engineering, it builds upon the foundation of existing research in automated ontology construction, knowledge representation, and semantic reasoning. A comprehensive review of prior work in these areas, including relevant methodologies and applications of LLMs in ontology engineering, is provided in Section 2. This ensures that our contributions are well grounded within the broader context of the existing literature.

The structure of the paper is as follows. Section 2 presents the related work. Section 3 describes the proposed approach. Section 4 presents the experiments and results, and Section 5 discusses the findings with respect to the research hypotheses. Finally, Section 6 summarizes the key findings of this study and identifies needs for future research.

2. Related Work

2.1. LLMs and OE

Zhang et al. [1] introduce a novel framework to streamline the process of OE (OE) through conversational interactions with a language model. Their study tackles the inherent complexities and resource-intensive nature of traditional OE, particularly in projects involving stakeholders from diverse backgrounds. OntoChat aims to mitigate systematic ambiguities and biases by facilitating requirement elicitation, analysis, and testing phases through a conversational AI interface. The framework enables users to create user stories and extract competency questions by interacting with a conversational agent, significantly reducing the manual effort typically required in these initial stages. The use of large language models (LLMs) in OntoChat allows for nuanced understanding and processing of natural language, enhancing the generation and refinement of ontology requirements. The paper details an evaluation of OntoChat using the Music Meta Ontology, demonstrating its effectiveness in improving the efficiency and accuracy of OE tasks. This approach not only streamlines the OE process but also promises to enhance the accessibility and adaptability of ontologies in various domains by incorporating advanced AI-driven methods.

Doumanas et al. [2] detail the evolution and application of OE methods in the context of large language models (LLMs), particularly focusing on human-to-machine centered methodologies. The paper discusses the gradual reduction in human involvement and corresponding increase in machine automation in the process of OE. This shift aims to leverage the efficiency of LLMs while retaining essential human oversight. The proposed spectrum of collaboration ranges from fully human-driven to completely LLM-driven OE methodologies. Through systematic experimentation across various levels of human and machine collaboration, the research evaluates the effectiveness of LLMs in creating ontologically sound structures with minimal human input. However, it is important to note that in this work, fine-tuning is not researched or evaluated; the study focuses solely on simple prompting techniques. The study provides a compelling look into how LLMs can significantly augment the OE process, making it faster and potentially more accurate while still needing human expertise for critical assessments and adjustments.

Garijo et al. [3] provide a detailed examination of how LLMs are applied in the domain of OE (OE). The paper categorizes various OE tasks that can benefit from LLMs, utilizing the Linked Open Terms (LOT) methodology as a framework for analysis. This research identifies the core phases of OE such as requirement specification, implementation, and maintenance, and highlights how LLMs are currently utilized within these areas. The authors discuss the diversity in task definitions and the need for standardized benchmarks and evaluation frameworks to measure LLM performance effectively in OE tasks. Through their analysis, Garijo et al. reveal significant gaps and suggest potential areas for further integration of LLMs to enhance the efficiency and effectiveness of OE processes.

Joachimiak et al. [4] discuss the development of the Artificial Intelligence Ontology (AIO), which organizes and defines AI concepts to support standardization and understanding in AI research. Developed at Lawrence Berkeley National Laboratory, the AIO employs LLMs to enhance manual curation, ensuring the ontology remains current with rapid advancements in AI technology. AIO is structured around six principal categories including Networks, Layers, and Bias, addressing both technical aspects and ethical considerations of AI. This approach not only facilitates modular composition of AI methodologies but also aids in navigating the ethical landscapes of AI applications, proving to be a vital resource for AI researchers and developers.

Saeedizade et al. [5] investigate the potential of LLMs to assist in the development of ontologies by generating OWL outputs directly from ontological requirements. The study explores several state-of-the-art models, utilizing a variety of prompting techniques such as Chain of Thoughts (CoT), Graph of Thoughts (GoT), and Decomposed Prompting to assess the ability of LLMs to produce sufficient quality OWL suggestions. The research demonstrates that GPT-4 is particularly capable of generating high-quality modeling suggestions, significantly outperforming other models in generating structurally and syntactically correct OWL files. This paper highlights the importance of carefully selected prompting techniques to leverage the capabilities of LLMs effectively in OE, suggesting a promising direction for automating this traditionally labor-intensive task.

Mateiu et al. [6] explore the potential of fine-tuning GPT-3 to automate the translation of natural language sentences into Description Logic, specifically into OWL Functional Syntax. Their research involves developing a Protégé plugin that assists in both developing new ontologies and enriching existing ones by automatically translating domain-specific natural language into formal OWL axioms under human supervision. The authors fine-tuned the GPT-3 model using a dataset of 150 natural languages to OWL translation pairs, covering a variety of instances and relationships. This approach leverages the linguistic prowess of LLMs to streamline the OE process, potentially reducing the technical challenges and costs traditionally associated with manual ontology development. Their work demonstrates a significant advancement in applying LLMs to semantic web technologies, particularly in automating and facilitating the labor-intensive processes of OE.

Doumanas et al. [7] explore the application of LLMs to OE within the context of Search and Rescue (SAR) operations. Their research demonstrates how LLMs can be effectively utilized to automate the construction of ontologies, particularly for complex and dynamic scenarios like SAR missions in wildfire incidents. However, it is important to note that in this work, fine-tuning is not researched or evaluated; the study focuses solely on simple prompting techniques. The paper proposes a novel collaborative OE methodology that harnesses both human expertise and the advanced natural language processing capabilities of LLMs. This hybrid approach facilitates the rapid development of domain-specific ontologies, leveraging the LLMs' ability to process and structure large datasets into ontological knowledge, which is critical for enhancing situational awareness and decision-making in SAR operations. The integration of LLMs into OE represents a significant advancement in making these operations more efficient and effective by providing a structured knowledge framework that supports the dynamic requirements of SAR missions.

2.2. LLMs and Fine-Tuning

Gekhman et al. [8] investigate the effects of supervised fine-tuning of large language models (LLMs) on the introduction of new factual knowledge. They explore whether such fine-tuning leads to an increase in the generation of factually incorrect responses—a phenomenon known as hallucination. The study reveals that LLMs tend to acquire new facts at a slower rate during fine-tuning, compared to facts that align with their pre-existing knowledge. Moreover, the findings indicate a linear relationship between the amount of new knowledge in the fine-tuning examples and the propensity of the model to hallucinate, highlighting the challenges of integrating new factual content into pre-trained models. The research underscores the importance of how LLMs are typically better at refining and utilizing their pre-existing knowledge rather than acquiring new information through fine-tuning. This has practical implications for the development and training strategies of LLMs, especially in ensuring that they remain reliable and factual in their outputs. It supports the notion that more controlled and carefully designed fine-tuning processes are required to enhance model reliability without compromising the accuracy and truthfulness of the content generated.

Chang et al. [9] propose a novel framework for enhancing the capabilities of pre-trained LLMs for time-series forecasting. Recognizing the inherent limitations of LLMs when applied to non-linguistic data, the study introduces a two-stage fine-tuning approach that adapts these models to the intricacies of time-series data. This includes a time-series alignment stage to familiarize the model with the time-series context, followed by a forecasting fine-tuning stage for specific forecasting tasks. Moreover, their method incorporates a unique two-level aggregation strategy to handle multi-scale temporal information effectively. The LLM4TS framework demonstrates superior performance across several datasets, outperforming state-of-the-art methods and offering significant improvements in few-shot scenarios, thus highlighting the potential of LLMs in domains beyond their initial linguistic training. This approach not only extends the applicability of LLMs but also enhances their efficiency and accuracy in handling complex time-series forecasting tasks.

Jeong [10] in his study delves into the nuances of employing LLMs specifically tailored for the financial sector. This research highlights the critical importance of dataset selection, preprocessing, and model choice, which are pivotal for fine-tuning LLMs to effectively address the unique challenges of financial data. By constructing domain-specific vocabularies and adhering to security and regulatory compliance, the study enhances the practical application of LLMs in financial services. Various financial scenarios, including stock price prediction, sentiment analysis of financial news, and automated document processing, illustrate the potential of fine-tuned LLMs to transform traditional financial operations into more efficient and insightful practices. Through detailed experimentation and analysis, Jeong contributes to the broader understanding of fine-tuning methodologies and their implications for the advancement of natural language processing technology within the business sector, specifically in finance.

Anisuzzaman et al. [11] discuss the methodology and impact of fine-tuning LLMs for domain-specific applications, particularly within the medical field. Their comprehensive review outlines the general steps and methodological approaches essential for effectively adapting pre-trained LLMs to specialized tasks. It highlights the use of models like ChatGPT and others for fine-tuning processes that significantly enhance their applicability in specific domains such as medical subspecialties. The paper illustrates how such fine-tuned models can assist in tasks ranging from pre-consultation and diagnosis to medical education and predictive analysis, ultimately aiming to improve accuracy and utility in practical healthcare settings. Furthermore, their study addresses the benefits and limitations associated with fine-tuning, emphasizing the importance of carefully managing the trade-offs between model customization and operational efficiency.

Raj J et al. [12] at HCLTech offer a detailed exploration of optimizing LLMs for enterprise applications. Their work focuses on the practical challenges and strategies for fine-tuning LLMs using proprietary enterprise data, highlighting the need for domain-specific adaptations to enhance performance and maintain data privacy. The authors provide a comprehensive guide on preparing data, selecting the right fine-tuning configurations, and utilizing advanced techniques like Low Rank Adaptation (LORA) and Quantized LORA for efficient training. They also discuss the use of retrieval-augmented generation (RAG) as an alternative to fine-tuning, which can leverage existing models with enhanced retrieval mechanisms to improve response quality without extensive retraining. This work is particularly valuable for organizations looking to implement LLMs effectively within their specific operational frameworks, ensuring that the models deliver high-quality, domain-relevant outputs while adhering to privacy and cost considerations.

Parthasarathy et al. [13] extensively explore the intricate processes and methodologies involved in fine-tuning LLMs. The report provides a detailed examination of various fine-tuning approaches, including supervised, unsupervised, and instruction-based methods, and delves into the challenges and opportunities each presents. It introduces a structured seven-stage pipeline for LLM fine-tuning, emphasizing critical stages such as data preparation, model initialization, and hyperparameter tuning. This detailed guide highlights innovative fine-tuning techniques like LoRA [14] and discusses the use of advanced configurations such as Mixture of Experts (MoE) [15] and Direct Preference Optimization (DPO) [16] to improve model performance and alignment with human preferences. Additionally, the study addresses the deployment of LLMs on distributed and cloud-based platforms, presenting a holistic view of the end-to-end fine-tuning process. This work serves as an essential resource for both researchers and practitioners, offering actionable insights into optimizing LLMs for specific applications while navigating the complexities of modern AI systems.

2.3. Summary

Leveraging LLMs for OE has demonstrated potential in automating and enhancing the OE process, yet it is underscored by critical limitations that suggest further research. Issues such as high computational demands and potential biases in training data, often derived from extensive web-crawled datasets, introduce data contamination and ethical concerns [1,2]. These issues highlight the essential need for continuous advancements in model architectures and training methodologies to elevate the reliability and fairness of LLM outputs. In the realm of OE, the integration of LLMs aims to streamline processes and decrease human labor, yet the importance of sustained human oversight to counteract biases and ensure the precision of generated ontologies is crucial [3–5]. Furthermore, the exploration of LLMs' automation capabilities in complex tasks like generating OWL outputs and translating natural language into Description Logic points out the challenges in achieving structurally and syntactically correct outputs [7,8]. Additionally, the research underscores the significant potential of LLMs in automating ontology construction for complex and dynamic scenarios like SAR operations, advocating collaborative methodologies that combine human expertise with LLMs' processing capabilities [9]. The need for domain-specific training and the integration of factual knowledge to mitigate the risk of generating incorrect responses is also emphasized [10]. Innovative frameworks that adapt LLMs to non-linguistic data further highlight the adaptability and broad applicability of these models [11]. Studies also point out the importance of tailored vocabularies and compliance with security regulations to enhance the effectiveness of LLMs in specific sectors like finance [12]. The benefits of fine-tuning LLMs for specialized medical tasks illustrate the models' utility in enhancing practical healthcare applications [13]. The strategic fine-tuning of LLMs using proprietary enterprise data showcases the necessity of domain-specific adaptations to maintain data privacy and performance [14]. Lastly, the need for a comprehensive guide on fine-tuning LLMs suggests the complexity of optimizing these models for various applications, advocating structured methodologies that address data preparation, model initialization, and performance enhancement [15].

This study aims to address these identified limitations, such as poor precision and recall and limited handling of domain-specific concepts, by implementing a comprehensive fine-tuning and evaluation strategy that enhances the capabilities of LLMs for OE while tackling the practical challenges highlighted in the related studies. Through the integration of advanced prompting techniques and iterative refinement processes, this methodology substantially improves the LLMs' understanding and generation of domain-specific ontologies, ensuring higher structural and syntactical accuracy. Rigorous human validation is also incorporated at critical stages of model training and output generation to mitigate biases and ensure the practical utility of the generated content. Furthermore, this approach leverages the computational efficiencies of LLMs but with a structured oversight mechanism that involves domain experts in the loop, ensuring that the ontologies produced are not only technically sound but also practically applicable. This framework contributes to the broader application of LLMs, establishing a more robust, adaptable, and scalable methodology for automating knowledge-intensive processes.

3. Research Methodology

3.1. Overview

The research presented in this paper focuses on fine-tuning two distinct LLMs, GPT-4 and Mistral 7B, to enhance their capabilities in generating ontologies. We selected GPT-4 and Mistral 7B for fine-tuning based on a combination of factors, including availability for fine-tuning, architectural differences, performance efficiency, and relevance to ontology engineering tasks. GPT-4, developed by OpenAI, is a high-parameter transformer model optimized for generating structured and semantically coherent outputs, making it well suited for handling ontology-related logical and syntactical requirements. It also offers robust reinforcement learning mechanisms that enhance its reasoning capabilities in complex domain-specific tasks. On the other hand, Mistral 7B was chosen as a lightweight alternative that prioritizes efficiency and cost-effectiveness. Its smaller architecture allows for faster inference and fine-tuning cycles, making it a viable option for scenarios where computational resources are constrained. The comparative assessment between these two models enables us to examine the trade-offs between accuracy, computational efficiency, and adaptability in ontology engineering applications.

The methodology encompasses several stages, starting from data preparation to the final evaluation of the models' performance. The objective is to equip these models with the necessary knowledge and accuracy in OE tasks, leveraging foundational texts to create domain-specific datasets.

3.2. Data Preparation

The first step involves creating datasets tailored for the fine-tuning process. This is achieved by selecting foundational texts in OE. These texts are thoroughly analyzed, and key concepts are extracted and transformed into question–answer pairs that align with OE requirements.

3.3. Creating Datasets

In our methodology for fine-tuning LLMs for OE, a pivotal goal was to automate the dataset creation process as much as possible. This automation is essential not just for improving efficiency, but also for ensuring consistency and scalability in the training datasets. By automating this process, we aimed to minimize human error and standardize the data quality across various training sets, which is crucial for the effective training of the models. The rationale behind automating dataset creation includes several key factors:

1. Scalability: Manual dataset creation is inherently labor-intensive and not scalable, particularly as the volume and complexity of source materials increase. Automation allows us to handle larger datasets efficiently, facilitating the development of models that can generalize across a broader spectrum of OE tasks.
2. Consistency: Manually created datasets can vary in quality and structure, depending on the individual's interpretation and method of data extraction. Automating this process with LLMs helps ensure a consistent format and quality across all datasets, thereby improving the reliability and effectiveness of the model training process.
3. Speed: Automation significantly accelerates the dataset creation process. LLMs can quickly process extensive texts, extracting relevant information and structuring it into the required format much faster than manual methods, thus saving valuable time and resources.
4. Resource Optimization: By automating routine data extraction and formatting tasks, valuable human resources can be redirected towards more complex and strategic tasks such as refining the models' architecture, configuring hyperparameters, and analyzing outcomes.

To implement automated dataset creation, the process involves several steps:

- Input and Segmentation: Selected foundational texts rich in OE content are fed into the LLMs, which automatically segmented them into manageable parts suitable for detailed processing.
- Extraction and Structuring: The LLMs are tasked with extracting essential concepts, definitions, and relational data from the texts. The extracted information is then automatically structured into question–answer pairs, formatted specifically to aid in effective model training.
- Validation and Refinement: Despite the automation, human oversight is essential. This phase involves validating and refining the LLM outputs to ensure that the data accurately reflects the required knowledge and is devoid of errors or irrelevant content.
- Output in JSONL Format: The structured datasets are then output in JSONL format (a collection of JSON values, where each line is a valid JSON value, typically an object or array), which is ideal for handling large volumes of data and simplifies integration into the model training workflows.

The selection of chapters from the foundational texts was guided by their relevance to core OE principles, OE methodologies and logical formalizations. We prioritized sections covering key topics such as ontology construction, class hierarchy definitions, object property constraints, and reasoning techniques. Chapters containing visual representations (diagrams, concept maps) were also selected to ensure that non-textual knowledge could be integrated into the dataset. These visual elements often encode relationships and taxonomies that are critical for domain representation but may not always be explicitly described in the text.

In more detail, diagrams in OE textbooks frequently illustrate relationships between concepts, taxonomical structures, and logical constrains. Since fine-tuning LLMs requires text-based datasets, it was necessary to convert these visual elements into a structured textual representation compatible with Q&A format. To systematically translate visual elements into description logic, we employed a multi-step approach: diagrams were analyzed to extract fundamental ontological elements such as classes, relationships and axioms. These extracted components were mapped to description logic expressions, and in case where ambiguities arose (e.g., conflicting relationships or missing formal constrains), textual descriptions from the same chapter were referenced for clarification. Any inconsistencies were manually reviewed and, when necessary, validated by domain experts to ensure logical coherence before incorporation into the dataset.

To ensure that the dataset represents a broad and comprehensive coverage of OE knowledge, we cross-validated the generated Q&A pairs against competency questions designed for ontology evaluation. Each selected chapter was assessed based on its contribution to foundational OE knowledge and its ability to support domain-specific ontology generation tasks. Furthermore, manual interventions were employed where automation introduced gaps-such as cases where multiple plausible interpretations of an ontology concept existed. These interventions ensured that the dataset maintained high accuracy and alignment with established OE principles.

Through this automated approach to dataset creation, we not only optimize the use of resources, but also enhance the models' training efficiency and efficacy. This sets a solid foundation for the subsequent stages of fine-tuning the LLMs, ultimately enabling them to generate practical, domain-specific ontologies effectively.

3.4. Model Fine-Tuning

The fine-tuning process is tailored for each model, with specific steps adjusted based on the characteristics and initial performance of GPT-4 and Mistral 7B.

3.4.1. GPT-4 Fine-Tuning

Fine-tuning GPT-4 for OE involves several strategic steps (Figure 1) designed to enhance its performance specifically for generating structured ontologies:

1. Initial Configuration: GPT-4 is set up with basic parameters tailored to ontology generation, ensuring it starts from a robust baseline.
2. Incremental Fine-Tuning: GPT-4 benefits from the ability to sequentially build upon previous fine-tunings. This means that each subsequent fine-tuning session starts from the last fine-tuned model, allowing the model to incrementally improve and adapt based on the cumulative knowledge it has acquired. This approach is efficient as it leverages prior adjustments without needing to retrain from scratch each time.
3. Parallel Prompt Processing: One of the unique capabilities of GPT-4 is its ability to run parallel prompts with GPT-4o and other pre-trained models from OpenAI. This feature is particularly useful in OE, as it allows the model to handle multiple aspects of ontology generation simultaneously, improving processing time and coherence in the generated outputs.
4. Iterative Refinement: The fine-tuning process is iterative, with ongoing adjustments based on intermediate results. This step is crucial to optimize the model's understanding of OE nuances and to refine its output quality.

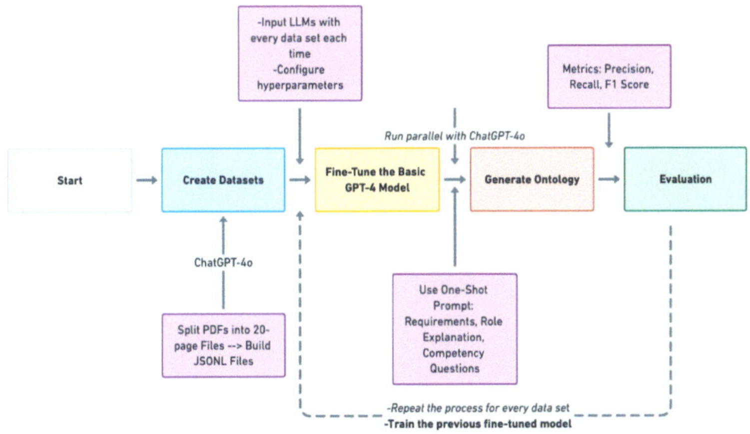

Figure 1. The process flow for fine-tuning the basic GPT-4 model for OE. The approach involves creating datasets from segmented PDFs, fine-tuning the model iteratively with domain-specific data, generating ontologies using a one-shoot prompt, and evaluating outputs through metrics.

3.4.2. Mistral 7b Fine-Tuning

Fine-tuning Mistral 7B, while similar in goal to GPT-4, involves different procedural steps (Figure 2) due to its distinct system capabilities:

1. Initial Setup: Each fine-tuning session with Mistral 7B starts with the basic Mistral model. Unlike GPT-4, Mistral does not support sequential fine-tuning on top of previously adjusted models. Therefore, each new fine-tuning session requires uploading not just the new dataset, but all previous datasets used in earlier sessions. This repetitive data loading is necessary to maintain continuity and consistency in the model's learning process.
2. Lack of Parallel Processing Capabilities: Mistral 7B does not have the capability to run parallel prompts or utilize concurrent processing with other pre-trained models. Each prompt must be processed sequentially. Initially, the basic Mistral model runs the prompt to establish a baseline before the fine-tuned versions are applied. This sequential approach ensures that each version of the model is strictly aligned with the incremental improvements intended through the fine-tuning.
3. Data Re-Upload for Each Session: For Mistral, each fine-tuning session is mainly standalone. This necessitates re-uploading all historical datasets for each new round of fine-tuning, a requirement that ensures the model does not lose the context or learning from previous sessions but does add to the setup time and computational load.

These detailed fine-tuning processes for GPT-4 and Mistral 7B showcase the nuanced approaches required to optimize each model's capabilities within the specific context of OE. GPT-4's ability to build upon previous enhancements and run parallel processing contrasts with Mistral 7B's need for repeated dataset uploads and sequential processing, illustrating the operational differences between these advanced LLM platforms.

In our previous research [7], we have presented the SimX-HCOME novel OE methodology, achieving substantial improvements in collaborative OE. Given these positive outcomes, we are reapplying SimX-HCOME to further refine and validate its effectiveness in automating and enhancing the OE process. This approach continues to leverage role-playing simulations among key stakeholders [17]—knowledge workers (KWs), domain experts (DEs), and knowledge engineers (KEs)—to streamline the development and verification of ontologies with reduced human intervention and increased accuracy.

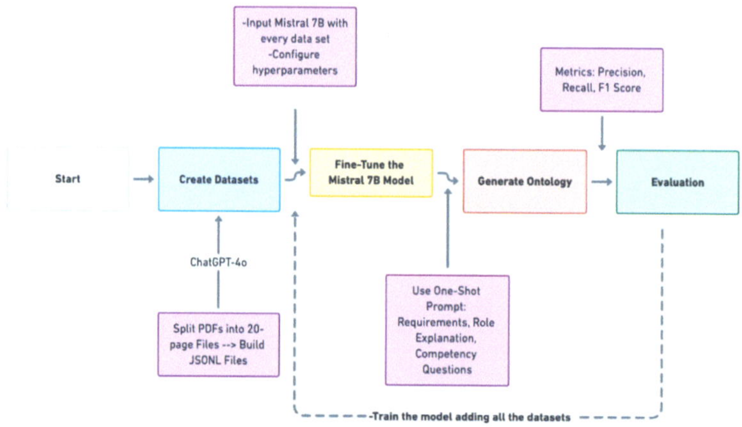

Figure 2. The process flow for fine-tuning the Mistral 7B model for OE. Each training session begins with the base model, requiring the aggregation of all datasets (new and previously used) for training. Continuation from the previously fine-tuned model is not supported, necessitating re-training from scratch with the combined datasets.

In the context of fine-tuning large language models (LLMs), such as GPT-4 (OpenAI Platform) and Mistral 7B (Mistral AI Le Platform), we opted to use an online platform that operates on a token-based payment system instead of conducting the fine-tuning locally. This decision was driven by several strategic considerations that align with the goals of accessibility, ease of use, and computational efficiency. Local fine-tuning requires substantial computational resources, often necessitating advanced GPUs, which can be prohibitively expensive and technically demanding. Online platforms mitigate this by offering access to state-of-the-art computing power on a pay-as-you-go basis, making it more cost-effective and avoiding the need for significant upfront investments in hardware. Moreover, these platforms are accessible to users without deep programming knowledge, simplifying the process through user-friendly interfaces and providing resources like pre-built scripts and comprehensive documentation. This democratizes access to advanced tools, allowing users from various backgrounds to experiment with and deploy AI solutions tailored to their needs. The community support and professional assistance available on these platforms also make them an invaluable resource for both novice and experienced users. Choosing to use an online platform with a token-based access model for fine-tuning LLMs strategically aligns with our objectives to maximize efficiency, reduce barriers to entry, and democratize access to cutting-edge AI technology, ensuring that individuals and organizations, regardless of their technical proficiency or resource availability, can harness the power of advanced LLMs for OE and other AI-driven endeavors.

To quantitatively assess the performance of our models, we compared their outputs against a reference ontology [18] constructed by human domain experts. This expert-created benchmark served as a gold standard, allowing us to evaluate precision, recall, and ultimately the F1 score of the generated ontologies. Our choice to focus on the domain of disaster management—particularly SAR operations in wildfire incidents—was not arbitrary. Members of our research team and co-authors possess direct professional experience in this area, having participated in real-world scenarios and policy development related to emergency response and resource allocation. Our prior research has already explored OE integrating LLMs in SAR missions. This existing domain knowledge makes it possible to ground our OE experiments in genuine, practice-informed requirements. By applying OE principles to a field we know intimately, we aim to ensure that the results are not only theoretically valid but also practically relevant.

This initial experiment thus lays a foundation for iterative improvements. Future steps will involve refining our approach to hyperparameters, expanding our training dataset, and incorporating iterative fine-tuning cycles. By doing so, we aim to enhance both the coverage and accuracy of generated ontologies and ultimately contribute meaningful, domain-specific insights to the field of disaster management and emergency response.

3.4.3. Computational Efficiency, Hardware Considerations, and Automated Hyperparameter Configuration

To ensure accessibility and ease of replication, we opted for pre-trained models rather than open-source alternatives that require specialized computational resources. This decision allows experiments to be conducted on standard computing devices without high-end hardware requirements. All experiments were performed on a MacBook Air M1 (2021) with 8 GB RAM, an 8-Core CPU, and a 7-Core GPU, demonstrating that our methodology does not rely on dedicated GPU clusters or cloud infrastructure. This choice ensures that our approach remains practical and reproducible for researchers working with limited resources.

To compare the computational efficiency of the models, we analyzed both fine-tuning time and cost. Mistral 7B required only three fine-tuning sessions, costing €4, €5.5, and €7 respectively, each taking approximately 5 min. In contrast, GPT-4 required an initial dataset loading phase lasting approximately one hour (as documented in our GitHub screenshot), with fine-tuning costs ranging from €7 to €10 per session, depending on dataset size. These results indicate that Mistral 7B offers a more cost-effective and time-efficient fine-tuning process, whereas GPT-4 requires additional time for dataset preparation but provides a more optimized API-based adaptation. These trade-offs should be considered when selecting a model for ontology engineering tasks.

In our fine-tuning experiments with both models, we opted for an automated hyperparameter configuration rather than manual tuning. This decision was motivated by the goal of making the fine-tuning process more accessible to researchers and practitioners without deep expertise in hyperparameter optimization. By leveraging the framework's default automated settings, the model was able to adjust learning parameters dynamically based on the dataset size, training steps, and optimization objectives, eliminating the need for manual intervention. Using automated hyperparameter settings led to a more stable training process but also introduced certain limitations. The model exhibited slower initial learning improvements, as the automated configuration did not include aggressive optimization strategies such as adaptive learning rate decay or manual gradient clipping. However, the benefit of this approach was that it reduced the risk of overfitting or instability, which can sometimes occur with aggressive hyperparameter tuning. Furthermore, the consistency observed in later fine-tuning iterations suggests that automated settings provided a balanced learning strategy, allowing Mistral 7B to generalize well across domain-specific ontology structures. While this approach facilitated a smoother fine-tuning experience, future research could explore the impact of alternative hyperparameter configurations, such as manually adjusting batch sizes, learning rates, or weight decay parameters, to further optimize model performance. Our methodology remains adaptable, allowing others to experiment with different settings while maintaining the overall workflow for fine-tuning LLMs in ontology engineering.

4. Experiments

4.1. Overview

The overarching goal of our work is to develop a fine-tuned model that can generate and curate structured semantic knowledge from carefully selected foundational texts. In this context, the experiments presented in this section aim to (i) substantiate the validity and effectiveness of the proposed approach, (ii) quantify the extent to which ontologies produced by the proposed method are accurate and complete and (iii) provide insight for the adjustment of hyperparameters and the fine-tuning cycles.

In the initial stage of our experimental framework, our primary aim was to tailor a state-of-the-art large language model (LLM)—specifically GPT-4—to meet the specialized demands of OE. By drawing upon authoritative works and core research in the domain, we sought to guide the model's internal representations toward not only understanding complex ontological concepts but also systematically distilling them into organized, readily accessible knowledge components.

Central to this effort was the translation of rich, often unstructured textual sources into a more structured question–answer (Q&A) format aligned with OE requirements. This reformatting step was intended to bring clarity and consistency to the knowledge extraction process, ensuring that the resultant information could be seamlessly integrated into ontology design and maintenance workflows. As stated above, the key focus was to produce model outputs that are both faithful to the source material and directly applicable to engineering tasks. In doing so, we established the foundation for subsequent experiments: refining the model's capabilities, assessing its performance against domain criteria, and ultimately advancing the state of the art in ontology-based systems development.

To ensure the model's training material was both authoritative and comprehensive, we began by selecting a curated set of foundational texts spanning the core principles, practices, and methodological frameworks of OE. These texts, specifically "Semantic Web for the Working Ontologist" by Dean Allemang and Jim Hendler [19], "A Semantic Web Primer" by Grigoris Antoniou, Paul Groth, Frank Van Harmelen, and Rinke Hoekstra [20], and "An Introduction to OE" by Maria Keet [21], were chosen for their well-established reputation and broad coverage of the field.

Drawing upon these works provided a rich, multifaceted source of knowledge that encompasses everything from basic terminology and conceptual underpinnings to advanced reasoning techniques and modeling strategies. Incorporating content from these references allowed us to capture the full spectrum of OE concerns—from high-level conceptual design to the practicalities of semantic data integration. By anchoring our data preparation workflow in these authoritative texts, we created a knowledge base that is both robust and aligned with established industry and research standards, ensuring that the fine-tuning process would build upon a solid intellectual foundation.

A crucial step in our data preparation process involved segmenting the source material into more granular, manageable units. Given that each of our selected foundational texts was structured into chapters typically spanning 20 to 30 pages, we chose the chapter as a natural starting point for segmentation. This segmentation was motivated both by practical considerations—such as the limits on the amount of text that could be processed effectively by the model at once—and by the desire to maintain thematic coherence within each unit of analysis. We aimed to ensure that each chapter's content was rich enough to provide meaningful context for knowledge extraction, yet not so large as to overwhelm the model's processing capacity.

To further refine this process, we adopted a working assumption that each page, on average, could yield approximately three to four Q&A pairs. Aggregated at the chapter level, this translated to roughly 60 Q&A pairs per chapter, providing a balanced dataset that captured a chapter's conceptual breadth without diluting its core insights. However, we soon encountered a key technical constraint: even at the chapter level, the complexity and length of the text could exceed GPT-4's input limitations. In response, we developed a strategy of breaking down chapters into smaller, more targeted passages before converting them into structured Q&A pairs. This approach not only ensured compliance with the model's constraints but also helped produce a data corpus that was both rich in detail and amenable to systematic fine-tuning.

To transform our collected textual and visual materials into a structured, machine-readable format, we employed a methodical approach centered on prompt engineering and iterative Q&A generation. Our overarching content request specified would produce a total of approximately 60 Q&A pairs per chapter to comprehensively capture all relevant knowledge. However, to maintain processing stability and manage the complexity of generating such a large number of pairs, we implemented a batching strategy. Instead of producing all 60 Q&A pairs in a single run, we limited each iteration to just 15 Q&A pairs, ensuring that the model could handle each request with minimal degradation in quality or coherence.

In addition to handling textual information, we also incorporated non-textual knowledge into this pipeline. Specifically, diagrams, schemas, and images found within the source material were not ignored; rather, we instructed the model to translate these visual artifacts into structured description logic forms. By doing so, we captured the structural and relational nuances that often cannot be conveyed through text alone. This step ensured that the final dataset not only reflected the factual and conceptual content of the chapters but also integrated their visual representations into a coherent semantic framework.

An illustrative example of our prompt design demonstrates the elaborate control exercised on the data generation process. For instance, consider a prompt asking for a total of 60 Q&A pairs derived from a chapter. The prompt might direct the model to first produce 20 fully formed, context-rich Q&A pairs—with "big-full responsive answers where is needed"—before proceeding to generate subsequent sets of Q&A pairs (Figure 3). Throughout the process, we emphasized adherence to the original chapter's structure and order, reminding the model to respect the natural progression of topics and ensure that no critical details were omitted or rearranged. Through careful prompt engineering and iterative, numerically controlled batching, we created a reliable pathway to building a high-quality JSONL dataset suitable for fine-tuning and subsequent OE tasks.

In order to ensure the generation of comprehensive and contextually accurate Q&A pairs, we employed a carefully crafted prompt. This prompt instructed the model to produce the questions and answers in batches, guided by the structure of the original text:

"I want in total 60 questions-answers. Give me now the first 20 messages-answers (big-full responsive answers where is needed) and remember when is necessary translate the visual content into descriptive text. You will continue to produce the rest of the messages-answers, after this prompt, so remember to give them in order of the chapter flow and not rearranged. Remember also you have to include all the knowledge on the pages and represent it in the queries".

Once the initial batches of Q&A pairs were generated, a systematic post-generation review process was conducted to ensure that each entry was both relevant and aligned with our OE objectives. During this evaluation, we identified and removed any generic or off-topic questions—such as those asking for a simple chapter summary without delving into underlying ontology concepts—because they did not enhance the model's conceptual understanding. This rigorous filtering step helped maintain the quality of our dataset by ensuring that each Q&A pair contributed substantively to the OE domain.

Following the initial pruning, we undertook a targeted refinement phase aimed at further enhancing the semantic richness and ontological depth of the Q&A pairs. Rather than settling for surface-level queries, we endeavored to guide our Q&A sets toward exploring ontology-related semantics, including classes, relations, and properties, as well as the use of RDF, OWL, and SPARQL. We also emphasized the integration of visual elements—originally presented as diagrams, schemas, or other graphical representations—into a formalized description logic framework. Through this iterative process, we progressively distilled the Q&A content into a resource that was not only high in quality and domain specificity, but also finely attuned to the principles and practices central to OE.

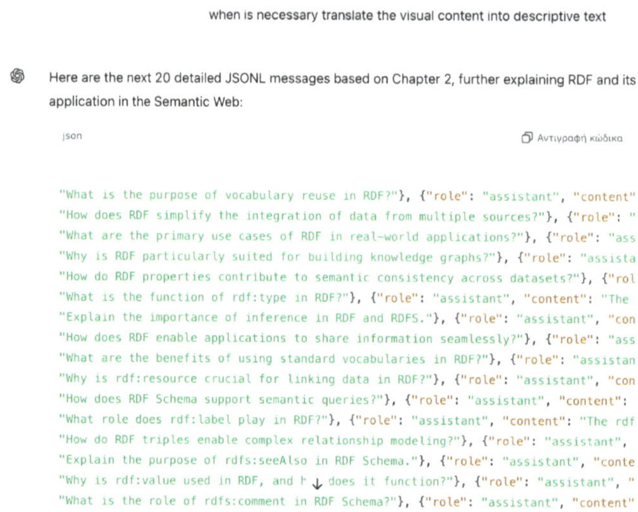

Figure 3. A screenshot from GPT-4 creating the first batch of 20 question–answer pairs for the datasets.

Throughout the data generation process, we encountered two main challenges. First, GPT-4 occasionally stopped responding before completing the requested number of Q&A pairs. To address this, we refined our prompting strategy by limiting each prompt to smaller batches of approximately 15 Q&A pairs. This approach reduced the cognitive load on the model, ensuring more stable and consistent outputs across iterations. The second challenge involved translating complex visual representations—such as diagrams and schemas—into coherent description logic statements. While the model could produce initial textual representations, often these required careful human verification to confirm accuracy and completeness. By manually reviewing and refining the model's outputs, we were able to ensure that the final Q&A sets faithfully captured both the textual and visual nuances of the source material.

4.2. Results

4.2.1. GPT-4 Fine-Tuning

As an initial step in our experimentation, we fine-tuned a GPT-4 model using a dataset derived exclusively from the book "Semantic Web for the Working Ontologist" by Dean Allemang and Jim Hendler [19]. This dataset, extracted and structured as Q&A pairs, aimed to capture the core ontological engineering principles, classes, relations, and foundational concepts detailed within the text. For this first fine-tuning run, we opted to rely on the automated configuration of hyperparameters to evaluate the out-of-the-box adaptability of the approach. Specifically, we trained the model for 3 epochs at a batch size of 1, utilized a learning rate (LR) multiplier of 2, and set the temperature to 1. After approximately 53 min, the fine-tuned model was ready for testing.

To thoroughly assess performance, we prompted both the newly fine-tuned model and a baseline GPT-4 version (here referenced as GPT-4o) with a scenario involving OE requirements, role-playing elements to simulate an iterative discussion, and a set of competency questions relevant to the domain. By running both models in parallel (Figure 4), we aimed to establish a comparative benchmark. The attached table (first row of the fine-tuning results—Tables 1 and 2) provides quantitative insights: While the fine-tuned model did not achieve the breadth of classes or properties identified by the reference standard, it did present different patterns of true positives, false positives, and false negatives compared to the basic GPT-4 baseline model. Although the fine-tuned model exhibited a higher precision—reflecting its tendency to produce fewer incorrect identifications—it came at the cost of lower recall, suggesting that it recognized fewer relevant concepts overall.

Table 1. Performance comparison of the baseline GPT-4 model and successive fine-tuned iterations in generating ontological classes, evaluated against a human expert reference ontology. The table highlights changes in true positives, false positives, false negatives, as well as in precision, recall, and F1 score, over multiple refinement steps.

Method	Number of Classes	True Positive	False Positives	False Negatives	Precision	Recall	F1 Score
Reference Ontology	80						
GPT-4 Basic Model	13	9	4	71	69%	11.25%	0.1935
GPT-4 Fine-Tuned Model 1	5	4	1	76	80%	5%	0.0941
GPT-4 Fine-Tuned Model 2	12	10	2	70	83%	12.5%	0.2173
GPT-4 Fine-Tuned Model 3	16	13	3	67	81.25%	16.25%	0.2708

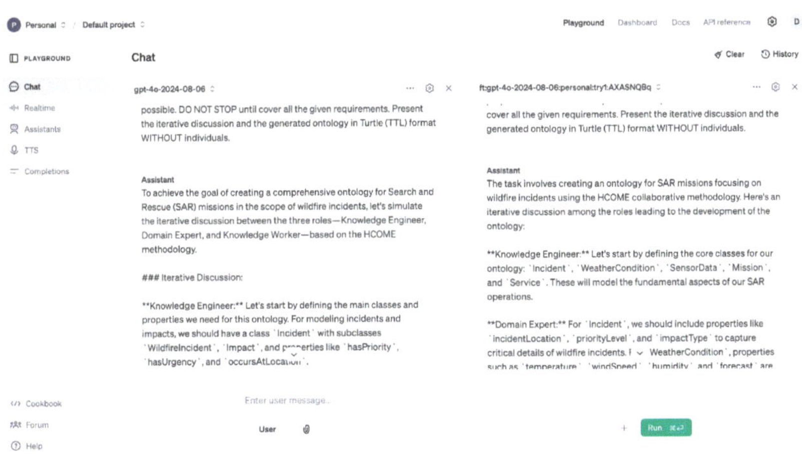

Figure 4. Running the same OE prompt side-by-side on both a baseline and a fine-tuned GPT-4 model, allowing for direct comparison of their role-based reasoning, conceptual coverage, and quality of generated ontologies.

Table 2. A summary of the GPT-4 model's performance on identifying and classifying object properties at various stages of fine-tuning, measured against a human expert reference ontology. The table provides true positives, false positives, false negatives, and the resulting precision, recall, and F1 scores for each iteration.

Method	Number of Obj. Properties	True Positive	False Positives	False Negatives	Precision	Recall	F1 Score
Reference Ontology	60						
GPT-4 Basic Model	5	0	5	60	0%	0%	0
GPT-4 Fine-Tuned Model 1	10	3	7	57	30%	5%	0.0857
GPT-4 Fine-Tuned Model 2	4	1	3	59	25%	1.66%	0.0312
GPT-4 Fine-Tuned Model 3	3	2	1	58	66.6%	3.33%	0.0634

In this early stage, the results highlight a critical tension in OE tasks: precision versus recall. The fine-tuned model, somewhat more conservative in its output, yielded fewer spurious results but also missed some relevant ones. These findings underscore the importance of iterative refinements and additional data augmentation. They also suggest that automated hyperparameter settings, while convenient, may not fully optimize performance for nuanced, domain-specific tasks, but that in our research is not the case as we try to automate the process as much as possible.

In the second fine-tuning iteration, we introduced additional training data from the book "A Semantic Web Primer" by Grigoris Antoniou, Paul Groth, Frank Van Harmelen, and Rinke Hoekstra [20], building upon the previously fine-tuned model trained solely on Semantic Web for the Working Ontologist. The same automated hyperparameters were retained: three epochs, a batch size of 1, an LR multiplier of 2, and a temperature of 1. This iteration completed in just 56 min, reflecting a relatively quick adaptation to the new knowledge.

When we re-ran the same OE prompt, the model once again assumed the prescribed role-playing format, producing an iterative conversation among the Knowledge Engineer, Domain Expert, and Knowledge Worker. This consistency in role adherence suggests that the foundational instructions from the first iteration had effectively "set the stage", allowing the model to integrate new conceptual material without losing the interactive structure.

Crucially, the results (Tables 1 and 2) indicate a marked improvement over the first iteration. Compared to the initial fine-tuned model, the second iteration revealed both an increase in true positives and a better balance between precision and recall. While the first iteration's fine-tuned model demonstrated high precision but struggled to identify a broad range of relevant concepts (resulting in low recall), the second iteration improved recall, capturing more of the classes and properties present in the reference ontology. As a consequence, the F1 score—an indicator of the model's overall effectiveness—rose substantially.

In practical terms, these enhancements mean that by supplementing the training data with content from a Semantic Web Primer, the model gained exposure to a broader spectrum of OE principles and examples. This enrichment allowed it to recognize and align with more elements of the expert-crafted reference ontology, thereby elevating its performance from merely being conservative and selective (high precision, low coverage) to a model that can identify a more representative set of relevant concepts (improved recall) while retaining strong precision. In short, the second fine-tuning iteration not only consolidated the model's grasp of established ontology concepts but also expanded its reach, making it more adept at reflecting the richness and diversity of a benchmark ontology constructed by human experts.

In the third fine-tuning iteration, we integrated an additional dataset derived from "An introduction to ontology engineering" by Maria Keet [21], supplementing the model's existing training from the two previous sources. Using the same automated hyperparameters (3 epochs, batch size of 1, LR multiplier of 2, and a temperature of 1), this fine-tuning process took approximately 52 min. Although the prompt remained structured around a three-role iterative discussion scenario, the model's response this time favored a more direct approach to ontology construction. It still recognized the roles, but was less dialog-oriented and more focused on systematically adding classes and properties to meet the competency requirements.

When compared to the previous two fine-tuning stages, the third iteration shows a notable evolution in the model's performance metrics (Tables 1 and 2). For classes, there is an uptick in both the number of classes correctly identified and the F1 score—a sign that the model is beginning to strike a more favorable balance between precision and recall. Earlier iterations either exhibited high precision but very low recall, or a slight improvement in recall at the cost of introducing more false positives. Now, with the introduction of concepts from Maria Keet's work, the model appears to have better internalized ontology design principles, allowing it to capture a broader spectrum of relevant classes while maintaining decent precision.

The results for object properties follow a similar pattern, though the improvements are more modest. Initially, the model struggled to identify many correct object properties, either missing the majority of them or producing extraneous, unrelated properties. After the third fine-tuning iteration, it identifies a slightly higher proportion of the correct object properties and does so with improved precision and a marginally better recall than before. While still far from exhaustive coverage, this incremental progress suggests that the model is starting to glean more nuanced relational structures from the domain literature.

The reasons behind these improvements likely stem from the cumulative knowledge gained: each additional training corpus deepens the model's understanding of ontological structures, methodologies, and best practices. The OE text by Keet, in particular, provides systematic insights into ontology design and evaluation, helping the model not only to propose classes and properties but also to do so in a manner more aligned with expert reference ontologies.

Nevertheless, the model's performance, while improved, still does not match the human expert reference. To achieve even better results, especially in terms of recall (covering more of the relevant concepts), it may be beneficial to introduce a targeted domain-specific dataset focused squarely on Search And Rescue missions in wildfire scenarios. Such a specialized dataset would provide richer domain knowledge, reduce guesswork, and help the model produce an ontology that is both semantically rigorous and thoroughly comprehensive. By grounding the model more firmly in the actual domain of disaster management and Search And Rescue operations, future fine-tuning efforts could yield an ontology that closely approaches the quality of expert-curated benchmarks (Further Experimentation Section).

4.2.2. Mistral 7B Fine-Tuning

For the Mistral 7B experiments, our methodology departed from the approach detailed in Section 3.4.1. While we continued to leverage the same foundational OE datasets, the operational workflow required a more hands-on, iterative process. This shift was necessitated by the distinct platform architecture and configuration requirements of the Mistral environment, which differ significantly from the streamlined setup provided by OpenAI. Unlike with the GPT-4 runs—where fine-tuning could be performed iteratively within the same ecosystem—adopting Mistral 7B involved re-uploading the entire dataset and re-establishing the training parameters each time a new fine-tuning run was initiated. This ensured a rigorously controlled refinement process but also introduced additional logistical overhead, as every iteration had to be executed from a fresh baseline tailored to the Mistral platform's unique constraints.

In this first fine-tuning iteration with Mistral 7B, we used the same OE dataset as before, but the process of adjusting hyperparameters was carried out manually because the platform did not have the option of automated selection. Despite the model successfully recognizing and maintaining the role-playing framework established in the prompt—distinguishing between a Knowledge Engineer, Domain Expert, and Knowledge Worker—it did not produce the requested ontology on the first attempt. Although the prompt explicitly instructed the model to generate the ontology, it initially provided only the iterative discussion without the final ontology construction. Consequently, we had to prompt it again, which resulted in the eventual creation of the ontology in Turtle (TTL) format (Figure 5).

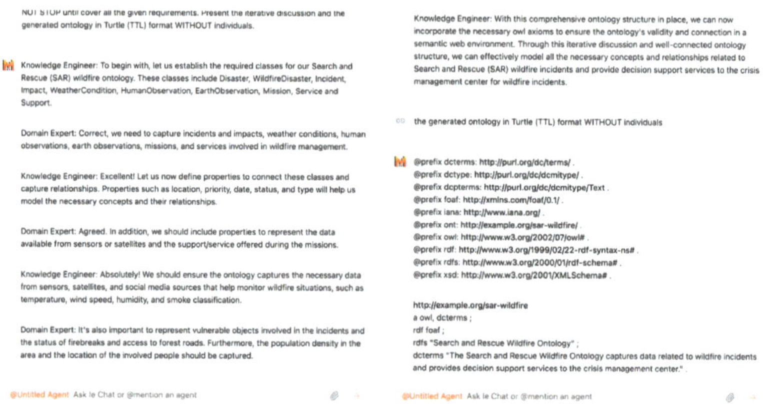

Figure 5. A side-by-side comparison of the baseline Mistral 7B model output (left) versus the fine-tuned version (right). Although the model understood the role-based scenario from the outset, it did not produce the requested ontology in response to the initial prompt and required an additional prompt before the ontology was finally generated in Turtle format.

Upon reviewing the ontology that Mistral 7B generated (Table 3), we observed that it produced a total of 21 classes. However, closer inspection revealed that 16 of these were duplicates, effectively leaving us with approximately 8 distinct classes. A plausible reason for this duplication could be that the model latched onto certain terms or phrase structures repeatedly, due to pattern replication rather than conceptual understanding. For instance, the model may have recognized the lexical pattern of a class name and recreated variations of it without introducing genuinely new semantic content. This behavior might stem from a combination of factors: the complexity of the prompt, the model's relative inexperience with the specific domain content (compared to more extensively fine-tuned models), or subtle ambiguities in the training data that led to repetitive outputs. Additionally, Mistral 7B, being a smaller model than GPT-4 and using a more manually handled fine-tuning process, may have fewer internal parameters to leverage for nuanced differentiation among closely related concepts. While the model did understand the narrative setup and roles, it struggled to follow through with the ontology generation on the first attempt and produced a partially redundant set of classes on the second. These initial results suggest that, at this stage, Mistral 7B may require more careful prompt engineering, possibly more domain-specific training data, or further iterative refinements to reduce redundancy and ensure that each generated class represents a truly distinct and meaningful ontological concept.

Table 3. The performance of the Mistral 7B model across various fine-tuning stages, compared to a human expert reference ontology. The table summarizes the number of classes identified, true positives, false positives, and false negatives, as well as the resulting precision, recall, and F1 scores for each iteration.

Method	Number of Classes	True Positive	False Positives	False Negatives	Precision	Recall	F1 Score
Reference ontology	80						
Mistral 7B Basic Model	19	14	6	67	68.4%	15.25%	0.2626
Mistral 7B Fine-tuned Model 1st	21	3	18	77	14.2%	3.75%	0.0594
Mistral 7B Fine-tuned Model 2nd	12	6	6	74	50%	7.5%	0.1304
Mistral 7B Fine-tuned Model 3rd	0	0	0	0	0%	0%	0

Compared to the Mistral Basic Model's initial performance on object properties, the first fine-tuned iteration presents a clear regression (Table 4). The baseline model identified some object properties correctly, registering a few true positives alongside several false positives. After the first fine-tuning, however, the model failed to produce any true positives and only produced false positives. In other words, it not only missed every correct object property from the reference set but also introduced erroneous ones. This result indicates that, rather than refining the model's understanding of relationships within the domain, the first round of fine-tuning appears to have destabilized its ability to discern object properties altogether. When placed in the context of previous approaches—such as the GPT-4 fine-tuned models, which at least maintained some level of correct object property identification—this drop is particularly concerning. GPT-4-based models, while imperfect, showed incremental gains in precision or recall after repeated fine-tuning. In contrast, the first fine-tuning of the Mistral 7B model led to a scenario in which the model no longer confidently identified any correct object properties, marking a significant performance gap compared to prior methods.

Table 4. The performance of the Mistral 7B model in identifying object properties at each fine-tuning stage, measured against a human expert reference ontology. The table details the progression (or regression) in terms of true positives, false positives, false negatives, and derived metrics (precision, recall, F1 score).

Method	Number of Obj. Properties	True Positive	False Positives	False Negatives	Precision	Recall	F1 Score
Reference ontology	60						
Mistral 7B Basic Model	17	4	13	56	23.5%	6.66%	0.1038
Mistral 7B Fine-tuned Model 1st	4	0	4	60	0%	0%	0
Mistral 7B Fine-tuned Model 2nd	0	0	0	0	0%	0%	0
Mistral 7B Fine-tuned Model 3rd	0	0	0	0	0%	0%	0

In this second fine-tuning iteration using Mistral 7B, we incorporated both the original dataset and the new one, effectively building upon the model that had already been exposed to the OE patterns from the first dataset. This cumulative training provided a stronger foundation of domain knowledge and formatting expectations. As a result, unlike in the first iteration, the model succeeded in generating the requested ontology on the first attempt, without needing a follow-up prompt.

Qualitatively, the model's response showed a more seamless integration of the role-based narrative and the final goal—producing a well-structured ontology (Table 3). The ontology's classes were introduced more coherently, reflecting a better conceptual mapping of key domain elements. While the quantitative metrics (precision, recall, and F1 score) may not have improved dramatically when compared to the previous iteration, the ability of the model to immediately produce a functioning ontology suggests it was better aligned with the task requirements. This indicates that the additional dataset and iterative fine-tuning have begun to guide the model toward a more stable and contextually appropriate representation of the ontology, making the second fine-tuning a noteworthy qualitative improvement over the first.

The second fine-tuning iteration did not improve upon the shortcomings observed after the first (Table 4). In fact, it eliminated all attempts at identifying object properties entirely—no true positives, but also no false positives. While removing false positives could be seen as the model becoming more conservative, the lack of any true positives or meaningful predictions indicates that the model has essentially "given up" on identifying object properties. This outcome might suggest that the additional training data or instructions did not help the model internalize the relationships between classes and their corresponding object properties. Instead of correcting course, the second fine-tuning iteration pushed the model toward a stance of total non-commitment. Comparing this to previous results, including the GPT-4 fine-tuned models, makes the problem stark. GPT-4 iterations typically attempted to refine their object property recognition, and even if they struggled, they did not default to a state of complete inactivity. The Mistral 7B model's second iteration shows no productive engagement with the domain's relational structures, falling well short of both its own baseline performance and GPT-4's incremental improvements in similar stages of refinement.

By the time we reached the third fine-tuning iteration with Mistral 7B, its performance in class identification had deteriorated to the point of producing no meaningful outputs at all (Table 3). This result starkly contrasts with the reference ontology's comprehensive structure of 80 classes and falls well short of any progress achieved in earlier attempts. In fact, while the first fine-tuning run offered some semblance of engagement—albeit with duplicates and low precision—and the second iteration at least attempted to identify a handful of correct concepts, the third yielded a complete absence of true positives. Compared to GPT-4's iterative refinements, which, despite challenges, consistently showed incremental gains or at least balanced trade-offs in precision and recall, Mistral 7B's final iteration shows no such adaptability. The model's stagnation could stem from multiple factors: its smaller size and capacity relative to GPT-4, the manual and more cumbersome re-upload process required for each new fine-tuning stage, and possible misalignments between the prompt instructions, data structure, and Mistral's internal representation capabilities. Taken together, these issues suggest that the model was neither fully assimilating the domain knowledge from the training sets nor adjusting its latent space to accurately reflect the reference ontology's complexity, ultimately resulting in disappointingly null outputs by the end of the third fine-tuning round.

Furthermore, the model's performance on object properties remained stagnant at zero across all relevant metrics: no true positives, no false positives, and no false negatives (Table 4). This flatline suggests that the model has not recovered or improved its internal representation of object properties. Instead, it remains in a neutral state of producing no object property identifications at all. In comparison, previous models—particularly the GPT-4-based ones—demonstrated some capacity for adaptation over multiple fine-tuning rounds, either increasing their precision or expanding their recall. The persistent zero-performance state in the Mistral model's third iteration emphasizes a critical failure to benefit from iterative training, and it underscores the challenge of aligning Mistral 7B with the complex relational reasoning required in OE tasks. While GPT-4 fine-tuning rounds often yielded at least incremental gains or trade-offs (such as improved precision at the cost of reduced recall), the Mistral 7B model's repeated failure to identify any object properties indicates that its training regime and data exposure might need substantial revision.

4.2.3. Error Analysis

To provide a deeper analysis of model performance, we examined errors in the generated ontologies to understand recurring limitations and potential challenged. Errors were primarily observed in three areas: (1) generating invalid axioms, where models produced statements that were either syntactically incorrect or logically inconsistent; (2) mismatched relationships, where object properties were incorrectly assigned, leading to incorrect domain and range specifications; and (3) poorly structured hierarchies, where models misclassified subclasses under incorrect parent concepts, affecting the overall ontology structure.

One notable example of invalid axiom generation was observed in Mistral 7B's handling of disjoint class constraints. It classified Firefighter as a subclass of FirstResponder, correctly indicating that all firefighters are first responders. However,, it then correctly classified Firefighter as a subclass of "not FirstResponder", implying that firefighters are explicitly not first responders. This contradiction creates an inconsistency in the ontology, as an entity cannot belong to both class and its negation. Such errors disrupt logical reasoning and can cause failures in automated inference systems. A correct ontology should either maintain Firefighter as a subclass of FirstResponder without contradiction or correctly define disjointness when necessary. Mistral 7B also incorrectly assigned the isAssignedTo relationship between firetrucks and firefighters. The model stated that a firetruck is assigned to a firefighter, which is incorrect. In reality, it is firefighters who are assigned to firetrucks, not the other way around. This error misrepresents the logical structure of the ontology and could lead to incorrect reasoning in applications. The correct relationship should define firefighters as the subject of the assignment, with firetrucks as the object. Poorly structured hierarchies were frequently observed in cases where models had to infer subclass relationships. For example, Mistral 7B placed "EmergencyResponseTeam" as a subclass of "FirstAidKit", which is a category error. GPT-4, on the other hand, successfully maintained the integrity of subclass structures but tended to create overly granular subclass divisions when unnecessary, leading to ontology bloat.

These observations highlight the different strengths and weaknesses of each model and reinforce the necessity of structured fine-tuning with domain-specific datasets to improve the logical coherence of ontology outputs.

4.3. *Discussion*

4.3.1. Quantitative Evaluation Metrics

The comparison between GPT-4 and Mistral 7B for generating domain-specific ontologies through fine-tuning reveals significant contrasts in their capabilities and highlights areas requiring further refinement. Both models were fine-tuned using foundational OE texts, and their performance was evaluated based on precision, recall, and F1 scores compared to a benchmark ontology crafted by human experts. The results from Section 4.2 illustrate the differing abilities of the two models to integrate and represent complex ontological knowledge.

GPT-4 demonstrated a notable capacity for incorporating foundational knowledge and iteratively improving its outputs through successive fine-tuning. Although the initial results were modest, each fine-tuning iteration led to incremental improvements in precision, such as reaching 83% in the second iteration, indicating better accuracy in identifying relevant classes. However, its recall metrics, such as 12.5% in the same iteration, reflect persistent challenges in capturing the breadth of the benchmark ontology. More specifically, for class identification, the precision of the basic model starts at 69% and improves to 83% by the second fine-tuned model, an increase of approximately 20.3%. In the final iteration, precision stabilizes at 81.25%, confirming a consistent upward trend. However, recall progresses more slowly, increasing from 11.25% in the basic model to 16.25% in the third fine-tuned model, a relative improvement of 44.4%. This discrepancy between precision and recall indicates that while GPT-4 becomes increasingly accurate, it still struggles to identify a broader set of relevant classes, reflected in its F1 score, which improves from 0.1935 in the basic model to 0.2708 in the final iteration—an overall improvement of approximately 39.9%. In the evaluation of object properties, GPT-4 begins with no true positives in its basic form, yielding a precision and recall of 0%. However, fine-tuning progressively improves precision, which reaches 66.6% by the third iteration. Despite this, recall remains extremely low, improving only marginally from 0% to 3.33%. As a result, the F1 score, which starts at 0, achieves only 0.0634 by the final iteration. GPT-4 consistently displayed a capacity to adapt and refine its understanding of the domain, effectively leveraging the prompts and incorporating iterative feedback.

In contrast, Mistral 7B faced more pronounced limitations. While the first fine-tuning iteration demonstrated the model's ability to generate classes, the presence of duplicate classes and a relatively low precision of 14.2% highlighted significant issues with internal consistency. Subsequent iterations showed slight improvements in precision and recall but remained far behind GPT-4 in both metrics. By the third iteration, Mistral 7B completely failed to produce meaningful outputs, with precision and recall dropping to 0%, indicating a complete breakdown in its ability to align with the benchmark ontology. These results underscore the inherent architectural and computational constraints of Mistral 7B compared to GPT-4. More specifically, for class identification, the basic model achieves a precision of 68.4% and a recall of 15.25%, resulting in an F1 score of 0.2626. However, the first fine-tuned iteration shows a sharp decline in performance, with precision dropping to 14.2%, a decrease of 79.2%, while recall plummets to 3.75%, reflecting a 75% reduction. Although precision recovers to 50% in the second fine-tuning iteration, recall remains low at 7.5%, and the final iteration fails entirely, producing no true positives and reducing recall to 0%. Despite an improvement in precision to 80% in the third iteration, this result is misleading, as it arises in the absence of any identified classes, effectively nullifying practical utility. In the case of object properties, Mistral 7B exhibits similar trends. The basic model begins with a precision of 23.5% and a recall of 6.66%, yielding an F1 score of 0.1038. However, fine-tuning results in complete failure across all iterations. The first fine-tuned model identifies no true positives, leading to a precision and recall of 0%. Subsequent iterations show no recovery, with precision and recall remaining at 0%. Even though the final iteration reports a precision of 80%, this figure is deceptive, as the absence of any true positives means the model fails to provide any meaningful object property identification.

The disparity in performance between the two models can be attributed to several factors. First, GPT-4's larger parameter space and advanced architecture equipped it to handle the logical reasoning and complexity required for OE tasks. Conversely, Mistral 7B's smaller size limited its ability to generalize and represent nuanced relationships effectively. Second, the fine-tuning methodologies differed significantly. GPT-4 employed an iterative approach, building upon previous fine-tuned models, which facilitated cumulative learning and enhanced performance. Mistral 7B, due to platform constraints, required re-uploading datasets for each iteration, disrupting continuity and hindering knowledge retention. Third, while GPT-4 consistently understood and executed the role-playing prompts designed to guide the ontology generation process, Mistral 7B struggled with prompt comprehension and produced incomplete or redundant outputs.

Despite GPT-4's superior performance, both models fell short of fully replicating the benchmark ontology. This highlights a critical limitation in their ability to generalize domain-specific knowledge solely from foundational texts. For instance, GPT-4 and Mistral 7B struggled with object properties, with the former achieving a maximum precision of 66.6% but low recall, while the latter failed to produce any meaningful outputs in later iterations. These shortcomings suggest a gap in the models' ability to integrate complex, domain-specific relationships and semantic structures.

To address these challenges, further experimentation is essential. Incorporating real-world domain-specific datasets, such as those derived from Search and Rescue (SAR) missions, could significantly enhance the models' contextual understanding and enable them to generate more robust and realistic ontologies. Such datasets would provide a richer foundation for fine-tuning, allowing the models to align more closely with practical requirements. Additionally, expanding the scope and depth of competency questions could help capture intricate relationships and guide the models toward improved precision and recall. This would not only refine their outputs but also make them more relevant to real-world applications.

In conclusion, while GPT-4 demonstrated greater potential for OE tasks compared to Mistral 7B, both models exhibited limitations that underscore the need for more comprehensive datasets and refined methodologies. Incorporating domain-specific data and leveraging iterative, feedback-driven approaches would likely yield more robust ontologies and bridge the gap between automated generation and expert-crafted benchmarks. These advancements could establish large language models as indispensable tools for domain-specific knowledge representation, ultimately enhancing their utility in OE and related fields.

4.3.2. Qualitative Analysis

Beyond numerical evaluation metrics, it is essential to analyze the ontologies' structural integrity, semantic coherence, and practical usability. The qualitative assessment provides deeper insight into the generated ontologies' logical consistency, hierarchy formation, and alignment with expected domain-specific knowledge.

1. Structural soundness: one of the key factors in assessing ontology quality is the correct hierarchical arrangement of concepts. GPT-4 consistently maintained structured and logically aligned hierarchies, preserving correct superclass-subclass (subsumption) relationships in most cases. However, it occasionally overgeneralized categories, grouping semantically distinct concepts under broader umbrella terms, requiring manual refinement. In contrast, Mistral 7B, particularly in early iterations, exhibited misclassifications and redundancy issues, such as assigning multiple conflicting parent classes to a single entity. However, with additional domain-specific fine-tuning, Mistral 7B showed significant improvement in class hierarchy consistency, correcting misplaced classifications and better differentiating subclass relationships.
2. Semantic coherence: the accuracy of relationships and constraints between entities is another important aspect of ontology quality. Mistral 7B struggled in initial fine-tuning iterations with incorrect domain and range assignments for object properties, often misaligning relationships (e.g., classifying an entity to an incorrect category). For example, it initially misclassified "EmergencyResponder" as both a subclass of

"MedicalPersonnel" and "Firefighter", leading to inconsistencies. After incorporating domain-specific datasets, the model demonstrated improved understanding, correctly placing "EmergencyResponder" as a broader category encompassing both roles without conflicting subclass assignments. GPT-4, on the other hand, rarely misclassified relationships but occasionally omitted necessary inverse relations, requiring manual correction.

3. Usability and practical implementation: a key consideration in evaluating ontology quality is its real-world applicability. GPT-4-generated ontologies were more structured and required minimal post-processing, making them suitable for immediate integration into knowledge-based systems. However, they sometimes lacked specificity, necessitating additional manual refinement. Mistral 7B required more adjustments in earlier iterations, but after targeted fine-tuning, its outputs became increasingly reliable, showcasing improved adaptability to domain-specific needs. This suggests that while GPT-4 provides a strong initial structure, Mistral 7B has the potential to generate highly customized ontologies when trained with more targeted data.

The evolution of Mistral 7B's ontology quality through iterative fine-tuning highlights the impact of dataset quality, competency question expansion, and domain adaptation. These refinements allowed the model to generate ontologies that better align with expert-defined structures. The comparison between GPT-4 and Mistral 7B demonstrates that while pre-trained models can produce well-formed structures, domain-specific fine-tuning plays a crucial role in refining their logical consistency and contextual accuracy.

4.3.3. Expert Evaluation on Ontology Quality

Beyond quantitative evaluation metrics, expert assessment plays a critical role in validating the logical coherence and practical applicability of generated ontologies. To further understand the strengths and weaknesses of the models, domain experts were consulted to assess the conceptual accuracy of class definitions and relationships, the logical consistency of generated axioms, and the overall utility of the ontology for real-world applications.

Experts noted that GPT-4 generally produced well-structured ontologies, but some inconsistencies appeared in domain-specific adaptations, requiring manual corrections. For example, in a generated ontology for Search and Rescue (SAR) operations, GPT-4 correctly classified IncidentCommander as a subclass of EmergencyResponder but mistakenly included it under both OperationalUnit and CommandUnit without clear distinction. This overlap had to be manually refined to maintain logical consistency.

Mistral 7B, on the other hand, was found to be more computationally efficient but required substantial post-processing due to missing or misclassified relationships. In some cases, it omitted necessary transitive relationships. For example, the model failed to infer that if a Firetruck isAssignedTo Station1 and Station1 isPartOf ResponseZoneA, then the Firetruck should be implicitly associated with ResponseZoneA. These inference gaps were commonly identified in expert reviews.

To ensure the reliability of expert assessments, multiple evaluators reviewed the ontologies independently. While formal inter-rater reliability calculations were not conducted, the consistency of independent expert feedback suggests a high level of agreement in assessing logical coherence and applicability. Experts particularly emphasized that while fine-tuned LLMs significantly aid ontology engineering, human validation remains essential to refine structure, resolve ambiguities, and correct logical inconsistencies.

4.4. Extended Experimentation

To enhance the robustness of the models and their ability to generate domain-specific ontologies, we expanded our experimentation by integrating two additional, domain-specific, datasets. These datasets were designed with the expertise of the authoring team, under their capacity as domain experts, and aimed to address the unique requirements of SAR operations during wildfire incidents. The need for domain-specific datasets beyond basic OE knowledge emerged from the findings of previous experiments, where the results fell short of representing the full capabilities of the large language models (LLMs) and their fine-tuning potential. While the models demonstrated some ability to process foundational OE concepts, their outputs lacked the contextual richness and applicability needed for complex, real-world scenarios. This highlighted the importance of integrating targeted, domain-specific datasets to boost the models' performance and enable them to generate more robust and nuanced ontologies.

The first dataset builds on the original set of 18 competency questions derived from the reference ontology. Recognizing gaps in coverage and the need for more granular representation of wildfire SAR operations, we subdivided the competency questions into subsections reflective of specific operational contexts as followed: Weather-Related Questions, Incident-Related Questions, Data from Human and Earth Observations, Missions and Services, Population and Community Impact, Environmental Impact and Rehabilitation, Technology and Modeling, Safety and Protocols and some Additional Questions. This structured expansion was guided by critical areas of knowledge that must be represented for effective SAR decision-making. By integrating domain expertise, we extended the dataset to 48 competency questions, ensuring comprehensive coverage of SAR ontology requirements. These competency questions were meticulously answered based on our domain expertise, forming a knowledge-rich dataset. The finalized Q&A pairs were uploaded into GPT-4 for JSONL conversion, ensuring consistency and accessibility for fine-tuning purposes. This expanded dataset aimed to provide nuanced insights and a broader perspective on SAR operations, enriching the model's training data and enabling it to represent complex, domain-specific scenarios accurately.

The second dataset was constructed from incident report templates used by the Greek Fire Department during actual wildfire incidents. These templates encapsulate critical information recorded during operations, such as resource allocation, mission timelines, environmental conditions, and outcomes. These real-world data were translated into a textual format and subsequently processed through GPT-4 to generate Q&A pairs in JSONL format. This dataset brings an operationally grounded perspective to the model, incorporating the terminology, procedural details, and contextual nuances encountered in SAR missions. By aligning the training data with practical applications, the dataset serves to bridge the gap between theoretical ontology principles and real-world use cases.

To evaluate the impact of these datasets, we fine-tuned the models in the configurations that each of them requires: on GPT-4 we followed the Separate Integration. Each dataset (expanded competency questions and incident reports) was uploaded independently to the last trained model (3rd fine-tuned) for fine-tuning. On Mistral 7B we comply with Sequential Integration. Following the sequence used in previous experiments, we fine-tuned the models first with the expanded competency questions dataset, followed by the incident report dataset. This method simulated a cumulative knowledge-building process, where the foundational ontology knowledge was augmented with real-world operational data.

Initial experiments suggest that incorporating these datasets leads to a noticeable improvement in the models' ability to generate robust, contextually accurate ontologies. The competency questions provided a theoretical framework, while the incident reports added operational depth, enabling the models to represent both strategic and tactical dimensions of SAR operations effectively. Further evaluation of precision, recall, and F1 scores, as well as qualitative assessments of generated ontologies, is ongoing to quantify these improvements. These datasets highlight the critical role of domain-specific knowledge in enhancing the utility of large language models for OE tasks, particularly in specialized applications like wildfire SAR missions.

4.4.1. GPT4

The new fine-tuned GPT-4 model demonstrated exceptional performance in generating a robust SAR ontology. By incorporating domain-specific datasets such as the expanded competency questions and wildfire incident reports, GPT-4 effectively utilized its role-playing capabilities to structure and generate a cohesive ontology. The model generated an ontology that aligns closely with practical SAR requirements, including detailed classifications of classes, object properties, and data properties (Tables 5 and 6). As seen in the ontology metrics, GPT-4 produced a structure with 72 distinct classes, 32 object properties, and 7 data properties, effectively capturing both theoretical and operational aspects. The model also achieved logical coherence, as evidenced by the 284 logical axioms and the robust use of subclass relationships. These results highlight GPT-4's ability to translate domain-specific knowledge into comprehensive and structured ontologies, showcasing its potential as a reliable tool for OE in specialized fields like wildfire SAR operations.

Table 5. A comparison of class generation metrics between the fine-tuned GPT-4 and Mistral 7B models after incorporating domain-specific datasets and their basic models, highlighting significant improvements in recall and F1 scores compared to previous experiments limited to OE training data.

Method	Number of Classes	True Positive	False Positives	False Negatives	Precision	Recall	F1 Score
Reference	80						
GPT-4 Basic Model	13	9	4	71	69%	11.25%	0.1935
Mistral 7B Basic Model	19	14	6	67	68.4%	15.25%	0.2626
New Fine-Tuned GPT-4	72	32	40	48	44.4%	40%	0.4210
New Fine-Tuned Mistral 7B	65	39	26	41	60%	48.75%	0.5379

Table 6. A comparison of object property generation metrics between the fine-tuned GPT-4 and Mistral 7B models after incorporating domain-specific datasets and their basic models, emphasizing significant improvements in recall and F1 scores over previous experiments relying solely on OE training data.

Method	Number of Obj. Properties	True Positive	False Positives	False Negatives	Precision	Recall	F1 Score
Reference	60						
GPT-4 Basic Model	5	0	5	60	0%	0%	0
Mistral 7B Basic Model	17	4	13	56	23.5%	6.66%	0.1038
New Fine-Tuned GPT-4	32	14	18	46	43.75%	23.33%	0.3043
New Fine-Tuned Mistral 7B	51	27	24	33	52.94%	45%	0.4864

More specifically, for class identification, the GPT-4 basic model achieved a precision of 69%, a recall of 11.25%, and an F1 score of 0.1935. In comparison, the new fine-tuned GPT-4 model demonstrated a precision of 44.4%, showing a 35.6% decrease, but its recall improved dramatically to 40%, marking a 255.5% improvement. This significant recall increase contributed to the F1 score rising to 0.4210, an overall improvement of 117.5%. For object property identification, the GPT-4 basic model started with 0% precision and 0% recall, producing an F1 score of 0. After fine-tuning, the new GPT-4 model achieved a precision of 43.75% and a recall of 23.33%, representing significant improvements over the baseline. The F1 score increased to 0.3043, highlighting the model's enhanced ability to identify object properties and maintain a reasonable balance between precision and recall. In summary, the new fine-tuned GPT-4 model demonstrates substantial gains over the basic model, particularly in recall, which shows remarkable improvement across both tasks. Despite a decrease in precision for class identification, the overall F1 score improvements confirm the effectiveness of the fine-tuning process.

4.4.2. Mistral 7B

The fine-tuned Mistral 7B model exhibited an impressive leap in performance during the generation of domain-specific ontologies for SAR operations. Unlike the earlier experiments, where Mistral 7B struggled compared to GPT-4, this iteration not only matched but surpassed GPT-4 in both depth and contextual accuracy. Leveraging domain-specific data, the model elevated the interaction among the three roles based on the HCOME OE methodology [17]. The iterative discussions, enriched by the model's ability to simulate nuanced domain knowledge, produced more insightful and well-rounded outputs, with enhanced clarity and purpose in defining and linking key concepts relevant to wildfire SAR scenarios. The ontology metrics further validate this success. Mistral 7B generated an ontology comprising 65 classes and 51 object properties, surpassing GPT-4 in capturing relationships and semantic depth (Tables 5 and 6). More specifically, for class identification, the Mistral 7B basic model achieved a precision of 68.4%, a recall of 15.25%, and an F1 score of 0.2626. After fine-tuning, precision decreased to 60%, reflecting a slight 12.3% decrease, but recall improved significantly to 48.75%, corresponding to an improvement equal to 219.7%. As a result, the F1 score rose to 0.5379, marking an overall improvement of 104.8%. This highlights a substantial enhancement in the model's ability to identify relevant classes, despite the slight decline in precision. For object property identification, the basic Mistral 7B model achieved a precision of 23.5%, a recall of 6.66%, and an F1 score of 0.1038. The new fine-tuned model improved precision to 52.94%, a 125.3% increase, and recall to 45%, reflecting an exceptional 575.7% improvement. This significant recall gain was the main contributor to the F1 score rising to 0.4864, an overall improvement of 368.4%.

Notable additions included classes and properties that detailed relationships between incidents, services, and environmental observations. The model introduced richer subclass hierarchies and properties, reflecting its improved grasp of the domain-specific complexities. This performance underscores the model's ability to semantically integrate knowledge, connecting heterogeneous data sources such as sensor observations and real-time weather conditions, and translating these into meaningful ontological structures. The fine-tuned Mistral 7B also excelled in leveraging discussions to optimize the ontology iteratively. The elevated level of role-play discussions facilitated not just the coverage of competency questions but also the generation of more context-aware and operationally useful ontologies. Mistral's ability to encapsulate operationally grounded and semantically rich outputs signifies its potential as a cost-efficient alternative to larger models, capable of yielding high-quality domain-specific ontologies tailored to real-world SAR challenges.

The performance of Mistral 7B exhibited significant inconsistencies across different fine-tuning sessions. Initial training attempts (as presented in Section 4.2.2, Tables 3 and 4) displayed notable deficiencies, such as class duplication, misclassification of object properties, and inconsistencies in relationship assignments. These issues primarily stemmed from early-stage hyperparameter settings, dataset preprocessing inconsistencies, and limitations in the initial prompt structures guiding the model. However, later iterations (Section 4.4.2, Tables 5 and 6) demonstrated a marked improvement, even surpassing GPT-4 in certain cases.

Several key factors contributed to this transition from suboptimal to high-performance results:

1. Data Preprocessing Adjustments: Early dataset formatting included overlapping or redundant entity descriptions, leading to class duplication and property assignment errors. By refining the dataset through more structured entity extraction and deduplication techniques, later training runs exhibited more consistent outputs.
2. Incorporation of Domain-Specific Datasets: A significant improvement was observed when ontology fine-tuning was expanded to include domain-specific datasets, which provided the model with richer contextual knowledge and helped mitigate early-stage errors in entity classification and relationship structuring. By grounding Mistral 7B in more domain-relevant examples, it was able to generate more coherent ontologies with improved logical consistency.
3. Expansion of Competency Questions: The number and complexity of competency questions used for model evaluation were increased in later iterations, providing a broader test set that reinforced correct interpretations of ontology structures. This iterative reinforcement led to improved model accuracy in reasoning-based ontology generation.
4. Iterative Fine-Tuning and Model Adaptation: The observed performance shift also highlights the cumulative effect of iterative fine-tuning, where each session builds upon prior refinements. While early fine-tuning attempts exposed model weaknesses, successive sessions reinforced correct classifications, improving coherence in ontology structuring.

These findings suggest that fine-tuning LLMs for ontology engineering requires a dynamic and adaptive strategy. Performance can fluctuate significantly based on domain-specific datasets and the iterative refinement process. This highlights the importance of continuous optimization and specific domain data to maximize model effectiveness in domain-specific knowledge representation.

5. Discussion

The results from Tables 5 and 6 underscore the transformative impact of incorporating domain-specific datasets into the fine-tuning process of large language models (LLMs), particularly in comparison to the earlier experiments outlined in Tables 1–4. These domain-specific datasets, focusing on wildfire SAR operations, allowed both GPT-4 and Mistral 7B to achieve significantly better precision, recall, and F1 scores, demonstrating their enhanced ability to represent domain-specific knowledge comprehensively.

In the earlier experiments (Table 1), GPT-4 exhibited limited recall and F1 scores even after iterative fine-tuning on basic OE knowledge. The baseline GPT-4 model achieved a precision of 69% but suffered from a recall as low as 11.25% and an F1 score of 0.1935. Similarly, Mistral 7B struggled with basic datasets, often failing to generate distinct classes without duplication, with minimal improvements across iterations.

With the introduction of domain-specific datasets, both models demonstrated significant performance gains in class generation. As shown in Table 5:

- GPT-4 improved its recall to 40%, doubling its performance compared to prior experiments. The addition of SAR-specific knowledge enriched the model's understanding, reflected in its ability to generate 72 distinct classes while maintaining a precision of 44.4%.

- Mistral 7B, previously outperformed by GPT-4 in class generation, surpassed GPT-4 in this iteration. It achieved a recall of 48.75% and an F1 score of 0.5379, emphasizing its ability to better integrate domain-specific insights into a semantically rich ontology with 65 classes. This represents a marked improvement over its earlier iterations (Table 3), where precision and recall were as low as 14.2% and 3.75%, respectively.

Object property generation results also highlight the value of domain-specific datasets (Table 6). Previously (Table 2), both GPT-4 and Mistral 7B struggled significantly with object property identification, with Mistral 7B failing to generate any true positives in later iterations and GPT-4 showing limited recall improvements.

- GPT-4, in its fine-tuned version with SAR datasets, generated 32 object properties, achieving a precision of 43.75% and an F1 score of 0.3043. Although this was a significant improvement over its earlier attempts (maximum recall of 3.33% in Table 2), it still lagged behind Mistral 7B in representing relational knowledge effectively.
- Mistral 7B, benefiting from sequential fine-tuning with domain-specific datasets, excelled in this category. It generated 51 object properties, achieving a recall of 45% and an F1 score of 0.4864. This improvement reflects the model's ability to map complex relationships, such as those between incidents, resources, and environmental factors, a crucial aspect of wildfire SAR operations.

The comparative analysis across these experiments underscores the critical role of domain-specific datasets in OE. While basic OE knowledge provides a foundational understanding, it falls short of enabling LLMs to address real-world, complex domains effectively. The SAR-specific datasets bridged this gap, enriching the models' knowledge bases with operationally grounded concepts and relationships.

The inclusion of expanded competency questions and incident report templates allowed the models to achieve the following:

- Better address the complexity of domain knowledge, including subclass hierarchies and property relationships.
- Improve recall, particularly in identifying classes and properties relevant to wildfire SAR operations.
- Enhance F1 scores, indicating a balanced ability to reduce false positives while capturing more true positives.

These results suggest that fine-tuning LLMs with targeted, domain-specific data are pivotal for achieving robust knowledge representation. Domain-specific datasets not only enable the models to better emulate human expert ontologies but also highlight their ability to capture operational nuances, such as environmental impacts, mission timelines, and resource allocation.

Furthermore, the performance disparity between Mistral 7B and GPT-4 in this context demonstrates that smaller models like Mistral, when equipped with rich domain-specific data, can outperform larger models trained on general datasets. This finding is particularly valuable for applications where computational resources or budget constraints favor smaller models.

Further experimentation reinforces the importance of aligning training datasets with practical, domain-specific requirements. By integrating such datasets, LLMs can achieve higher precision and recall, bridging the gap between theoretical OE and real-world applications. These insights pave the way for more refined methodologies and hybrid approaches to optimize knowledge representation in specialized fields like wildfire SAR operations.

6. Conclusions

This research paper explores the fine-tuning of LLMs, specifically GPT-4 and Mistral 7B, for OE tasks. Through comprehensive experiments, the impact of general OE training and domain-specific fine-tuning on the models' ability to generate robust and practical ontologies was evaluated. The results validated the identified research questions and revealed critical insights into the role of targeted data in enhancing the models' performance.

The first research question (RQ1) related to the fine-tuning of LLMs on general OE concepts and the improvement in their performance was partially validated. While fine-tuning improved the baseline capabilities of both GPT-4 and Mistral 7B, the results highlighted a few limitations. GPT-4 demonstrated incremental improvements, achieving a precision of 81.25% and a modest recall increase from 11.25% to 16.25% in OWL class generation after three iterations. Mistral 7B, however, struggled with basic OE datasets, showing limited progress and regression in key metrics, with a recall of only 15.25% in preliminary experiments. These findings suggest that while foundational OE training is essential, it alone cannot shape models for the nuanced requirements of domain-specific knowledge representation.

The second research question (RQ2)—that incorporating domain-specific datasets would enhance the practical utility of generated ontologies—was strongly validated. The addition of SAR-specific datasets, including expanded competency questions and real-world wildfire incident reports, significantly improved the models' performance. GPT-4's recall for class generation increased to 40%, with an F1 score of 0.4210, while its recall for object properties rose to 23.33%. More notably, Mistral 7B, which had previously lagged, surpassed GPT-4 in both class and object property metrics. It achieved a recall of 48.75% and an F1 score of 0.5379 for classes, and a recall of 45% and an F1 score of 0.4864 for object properties. These results underscore the transformative impact of domain-specific data in enabling models to accurately represent complex and operationally grounded knowledge structures.

The key findings from this study highlight the necessity of combining domain-specific datasets and fine-tuning methodologies for maximizing LLM performance in specialized tasks. The integration of SAR-specific data enriched the models' understanding of key relationships and hierarchies, enabling the generation of semantically rigorous and practically relevant ontologies. Mistral 7B's performance in these experiments was particularly noteworthy, as it not only overcame its earlier struggles but also outperformed GPT-4 in recall and F1 scores for both classes and object properties. This demonstrates that smaller, cost-efficient models can achieve competitive results when trained on targeted data.

This paper also revealed the importance of aligning training data with real-world applications. Earlier experiments using general OE data showed limited recall and low F1 scores, indicating a lack of contextual understanding. In contrast, domain-specific fine-tuning improved both models' ability to generate ontologies that aligned with practical needs, including detailed representations of SAR concepts such as weather impacts, resource allocation, and incident timelines. These improvements emphasize that domain-specific datasets bridge the gap between theoretical ontology principles and operational utility. In conclusion, this research confirms the critical role of fine-tuning with domain-specific datasets in enhancing the utilization of LLMs for OE. While general OE training lays the groundwork, it is insufficient for generating robust, application-ready ontologies. Domain-specific datasets enable models to achieve significantly higher recall and F1 scores, reflecting better contextual understanding and practical relevance. The findings demonstrate the potential of LLMs, particularly when fine-tuned with targeted data, to advance OE and serve as indispensable tools for knowledge representation in specialized fields such as wildfire SAR operations. These insights pave the way for future studies to refine fine-tuning methodologies and extend the applicability of LLMs across diverse domains.

While this study demonstrates the feasibility of fine-tuning LLMs for ontology engineering in SAR contexts, an important next step is the integration of these ontologies into operational decision support systems. In practical applications, ontology-driven reasoning can be embedded into SAR platforms to enhance incident classification, automate information retrieval, and support knowledge-based decision-making. For example, ontology-enhanced knowledge graphs could improve real-time situational awareness by dynamically linking incident reports, geographical data, and response protocols to assist emergency teams.

Future improvements could involve incorporating retrieval-augmented generation (RAG) strategies and active learning approaches to continuously refine the model based on real-time domain expert feedback. This would allow the system to adapt to new terminologies, evolving threat landscapes, and emerging response protocols in SAR environments. Such an approach could ensure that the model remains aligned with the operational needs of first responders while improving its generalization capabilities across different emergency scenarios.

Furthermore, we recognize the need for additional details regarding data preparation steps, computational efficiency, and real-world deployment scenarios. Future research could explore a fully operational prototype of an ontology-powered SAR assistant, systematically evaluating scalability, response accuracy, and integration challenges in real-world settings. These enhancements would further solidify the role of LLMs in ontology-driven decision support at scale.

The experiments detailed (source files, screenshots, ttl files and chat conversations) are saved in the following GitHub repository: https://github.com/dimitrisdoumanas19/Fine-tuning-LLMS.git (accessed on 7 December 2024).

Author Contributions: Conceptualization, D.D., A.S. and K.K.; Methodology, D.D., A.S. and K.K.; Writing—original draft, D.D.; Writing—review & editing, D.D., D.S., C.V. and K.K.; Supervision, D.S., C.V. and K.K. All authors have read and agreed to the published version of the manuscript.

Funding: This research received no external funding.

Institutional Review Board Statement: Not applicable.

Informed Consent Statement: Not applicable.

Data Availability Statement: The data presented in this study are openly available in https://github.com/dimitrisdoumanas19/Fine-tuning-LLMS.git, accessed on 7 December 2024.

Conflicts of Interest: The authors declare no conflict of interest.

References

1. Zhang, B.; Carriero, V.A.; Schreiberhuber, K.; Tsaneva, S.; González, L.S.; Kim, J.; de Berardinis, J. OntoChat: A Framework for Conversational Ontology Engineering using Language Models. *arXiv* **2024**, arXiv:2403.05921. [CrossRef]
2. Doumanas, D.; Soularidis, A.; Kotis, K.; Vouros, G. Integrating LLMs in the Engineering of a SAR Ontology. In *Artificial Intelligence Applications and Innovations;* Maglogiannis, I., Iliadis, L., Macintyre, J., Avlonitis, M., Papaleonidas, A., Eds.; Springer Nature: Cham, Switzerland, 2024; pp. 360–374. [CrossRef]
3. Garijo, D.; Poveda-Villalón, M.; Amador-Domínguez, E.; Wang, Z.; García-Castro, R.; Corcho, O. LLMs for Ontology Engineering: A Landscape of Tasks and Benchmarking Challenges. In Proceedings of the International Semantic Web Conference ISWC2024, Baltimore, MD, USA, 11–15 November 2024.
4. Joachimiak, M.P.; Miller, M.A.; Caufield, J.H.; Ly, R.; Harris, N.L.; Tritt, A.; Mungall, C.J.; Bouchard, K.E. The Artificial Intelligence Ontology: LLM-assisted construction of AI concept hierarchies. *arXiv* **2024**, arXiv:2404.03044.

5. Saeedizade, M.J.; Blomqvist, E. Navigating Ontology Development with Large Language Models. In *The Semantic Web*; Lecture Notes in Computer Science; Peñuela, A.M., Dimou, A., Troncy, R., Hartig, O., Acosta, M., Alam, M., Paulheim, H., Lisena, P., Eds.; Springer Nature: Cham, Switzerland, 2024; Volume 14664, pp. 143–161. [CrossRef]
6. Mateiu, P.; Groza, A. Ontology engineering with Large Language Models. In Proceedings of the 2023 25th International Symposium on Symbolic and Numeric Algorithms for Scientific Computing (SYNASC), Nancy, France, 11–14 September 2023.
7. Doumanas, D.; Bouchouras, G.; Soularidis, A.; Kotis, K.; Vouros, G. From Human- to LLM-centered Collaborative Ontology Engineering. Large Language Models (LLMs) and Ontologies. *Appl. Ontol.* **2024**, 15705838241305067.
8. Gekhman, Z.; Yona, G.; Aharoni, R.; Eyal, M.; Feder, A.; Reichart, R.; Herzig, J. Does Fine-Tuning LLMs on New Knowledge Encourage Hallucinations? *arXiv* **2024**, arXiv:2405.05904. [CrossRef]
9. Chang, C.; Wang, W.-Y.; Peng, W.-C.; Chen, T.-F. LLM4TS: Aligning Pre-Trained LLMs as Data-Efficient Time-Series Forecasters. *arXiv* **2023**, arXiv:2308.08469. [CrossRef]
10. Jeong, C. Fine-tuning and Utilization Methods of Domain-specific LLMs. *arXiv* **2024**, arXiv:2401.02981.
11. Anisuzzaman, D.M.; Malins, J.G.; Friedman, P.A.; Attia, Z.I. Fine-Tuning LLMs for Specialized Use Cases. *Mayo Clin. Proc. Digit. Health* **2024**, *3*, 100184. [CrossRef]
12. J, M.R.; VM, K.; Warrier, H.; Gupta, Y. Fine Tuning LLM for Enterprise: Practical Guidelines and Recommendations. *arXiv* **2024**, arXiv:2404.10779. [CrossRef]
13. Parthasarathy, V.B.; Zafar, A.; Khan, A.; Shahid, A. The Ultimate Guide to Fine-Tuning LLMs from Basics to Breakthroughs: An Exhaustive Review of Technologies, Research, Best Practices, Applied Research Challenges and Opportunities. *arXiv* **2024**, arXiv:2408.13296. [CrossRef]
14. Pathak, A.; Shree, O.; Agarwal, M.; Sarkar, S.D.; Tiwary, A. Performance Analysis of LoRA Finetuning Llama-2. In Proceedings of the 2023 7th International Conference on Electronics, Materials Engineering & Nano-Technology (IEMENTech), Kolkata, India, 18–20 December 2023; pp. 1–4. [CrossRef]
15. Nguyen, H.D.; Chamroukhi, F. Practical and theoretical aspects of mixture-of-experts modeling: An overview. *WIREs Data Min. Knowl. Discov.* **2018**, *8*, e1246. [CrossRef]
16. Bernardelle, P.; Demartini, G. Optimizing LLMs with Direct Preferences: A Data Efficiency Perspective. In Proceedings of the 2024 Annual International ACM SIGIR Conference on Research and Development in Information Retrieval in the Asia Pacific Region, Tokyo, Japan, 9–12 December 2024; pp. 236–240. [CrossRef]
17. Kotis, K.; Vouros, G.A. Human-centered ontology engineering: The HCOME methodology. *Knowl. Inf. Syst.* **2006**, *10*, 109–131. [CrossRef]
18. Masa, P.; Meditskos, G.; Kintzios, S.; Vrochidis, S.; Kompatsiaris, I. Ontology-based Modelling and Reasoning for Forest Fire Emergencies in Resilient Societies. In Proceedings of the 12th Hellenic Conference on Artificial Intelligence, Corfu, Greece, 7–9 September 2022; pp. 1–9. [CrossRef]
19. Allemang, D.; Hendler, J. *Semantic Web for the Working Ontologist*; Elsevier: Amsterdam, The Netherlands, 2011. [CrossRef]
20. Antoniou, G.; Groth, P.; van Harmelen, F.; Hoekstra, R. A Semantic Web Primer. In *Cooperative Information Systems*, 3rd ed.; MIT Press: Cambridge, MA, USA, 2012.
21. Keet, C.M. *An Introduction to Ontology Engineering*; College Publications: London, UK, 2018.

applied sciences

Article

MCP: A Named Entity Recognition Method for Shearer Maintenance Based on Multi-Level Clue-Guided Prompt Learning

Xiangang Cao [1,2], Luyang Shi [1,2,*], Xulong Wang [1,2], Yong Duan [1,2], Xin Yang [1,2] and Xinyuan Zhang [1,2]

1 School of Mechanical Engineering, Xi'an University of Science and Technology, Xi'an 710054, China; cao_xust@sina.com (X.C.); 15838959020@163.com (X.W.); duanyong@stu.xust.edu.cn (Y.D.); 21105016014@stu.xust.edu.cn (X.Y.); 13679172944@163.com (X.Z.)
2 Shaanxi Key Laboratory of Mine Electromechanical Equipment Intelligent Monitoring, Xi'an 710054, China
* Correspondence: 23205016034@stu.xust.edu.cn

Abstract: The coal mining industry has accumulated a vast amount of knowledge on shearer accident analysis and handling during its development. Accurately identifying and extracting entity information related to shearer maintenance is crucial for advancing downstream tasks in intelligent shearer operations and maintenance. Currently, named entity recognition in the field of shearer maintenance primarily relies on fine-tuning-based methods; however, a gap exists between pretraining and downstream tasks. In this paper, we introduce prompt learning and large language models (LLMs), proposing a named entity recognition method for shearer maintenance based on multi-level clue-guided prompt learning (MCP). This method consists of three key components: (1) the prompt learning layer, which encapsulates the information to be identified and forms multi-level sub-clues into structured prompts based on a predefined format; (2) the LLM layer, which employs a decoder-only architecture-based large language model to deeply process the connection between the structured prompts and the information to be identified through multiple stacked decoder layers; and (3) the answer layer, which maps the output of the LLM layer to a structured label space via a parser to obtain the recognition results of structured named entities in the shearer maintenance domain. By designing multi-level sub-clues, MCP enables the model to extract and learn trigger words related to entity recognition from the prompts, acquiring context-aware prompt tokens. This allows the model to make accurate predictions, bridging the gap between fine-tuning and pretraining while eliminating the reliance on labeled data for fine-tuning. Validation was conducted on a self-constructed knowledge corpus in the shearer maintenance domain. Experimental results demonstrate that the proposed method outperforms mainstream baseline models in the field of shearer maintenance.

Keywords: shearer; large language models; prompt learning; named entity recognition

1. Introduction

As one of the key pieces of equipment in fully mechanized mining systems, the shearer's efficient and stable operation is critical for ensuring coal mine production efficiency, guaranteeing safe operations, and reducing maintenance costs. Various multi-source data, such as shearer maintenance manuals and professional books, serve as vital knowledge sources for extracting and summarizing maintenance concepts and practical experience. Accurately identifying and extracting diverse entity information related to shearer maintenance is an essential prerequisite for enabling intelligent operation and maintenance downstream tasks such as knowledge graphs, question-answering systems, and information retrieval. It plays an irreplaceable role in the intelligent framework of coal production processes. Therefore, there is an urgent need to develop efficient named entity recognition (NER) methods in the field of shearer maintenance.

NER refers to the process of identifying entities with specific meanings in a text and categorizing them into predefined entity types. Early NER methods primarily relied on expert-designed rules and templates to identify named entities [1]. However, these approaches had limitations, including high rule-development costs and poor transferability to other corpora [2]. With the rise of machine learning in the field of natural language processing (NLP), researchers have explored NER using machine learning methods, such as hidden Markov models (HMMs) [3], conditional random fields (CRFs) [4], maximum entropy models (MEMs) [5], and support vector machines (SVMs) [6]. These methods rely on learning features and patterns of input text from large corpora and utilize feature engineering to extract morphological characteristics of the input text to achieve NER. However, machine learning methods often require manually designed features and exhibit limited generalization capability.

With the continuous development of deep learning technology, deep learning-based methods typically leverage large-scale, high-quality annotated corpora to train complex network models, resulting in better entity recognition performance compared to machine learning methods. For example, various sequence labeling models based on LSTM [7], such as LSTM, Bi-LSTM, and Bi-LSTM-CRF, have been proposed. In recent years, with the emergence of pretrained language models (PLMs), the bidirectional encoder representations from transformer (BERT) model [8], based on transformer architecture, has significantly enhanced the understanding of syntactic and semantic information, bringing substantial improvements to various NLP tasks. The advent of pretrained models has driven a new research paradigm known as "fine-tuning". Fine-tuning refers to using PLMs as backbone models and performing targeted training on some or all of their parameters using labeled target datasets for specific downstream tasks. Researchers construct models by appending network structures specifically designed for downstream tasks to the end of PLMs. NER in the shearer maintenance domain primarily employs methods based on the fine-tuning paradigm. As shown in Figure 1a,b, during the fine-tuning process, the appended network models may introduce structural inconsistencies, and the differing objectives of upstream and downstream tasks further constrain their performance. Moreover, these methods heavily depend on the quality and size of labeled datasets.

Recently, some studies have proposed a novel approach called prompt learning [9,10]. This method bridges the gap between pretraining and downstream tasks by converting various downstream tasks into optimal cues or front-end prompts, such as cloze tests. As shown in Figure 1c, this is an example of a prompt learning method. This paradigm integrates the original input with a prompt template containing "[MASK]", then feeds the combined sequence into the language model to predict the "[MASK]". Generative LLMs like ChatGPT [11,12] have demonstrated outstanding performance in tasks such as dialogue, text understanding, and reasoning. Extensive studies have shown that ChatGPT outperforms zero-shot learning models in most NLP tasks and even surpasses fine-tuned models in certain tasks. Furthermore, it has demonstrated remarkable capabilities in low-resource NLP tasks, marking a transition in NLP tasks from traditional PLMs to LLMs.

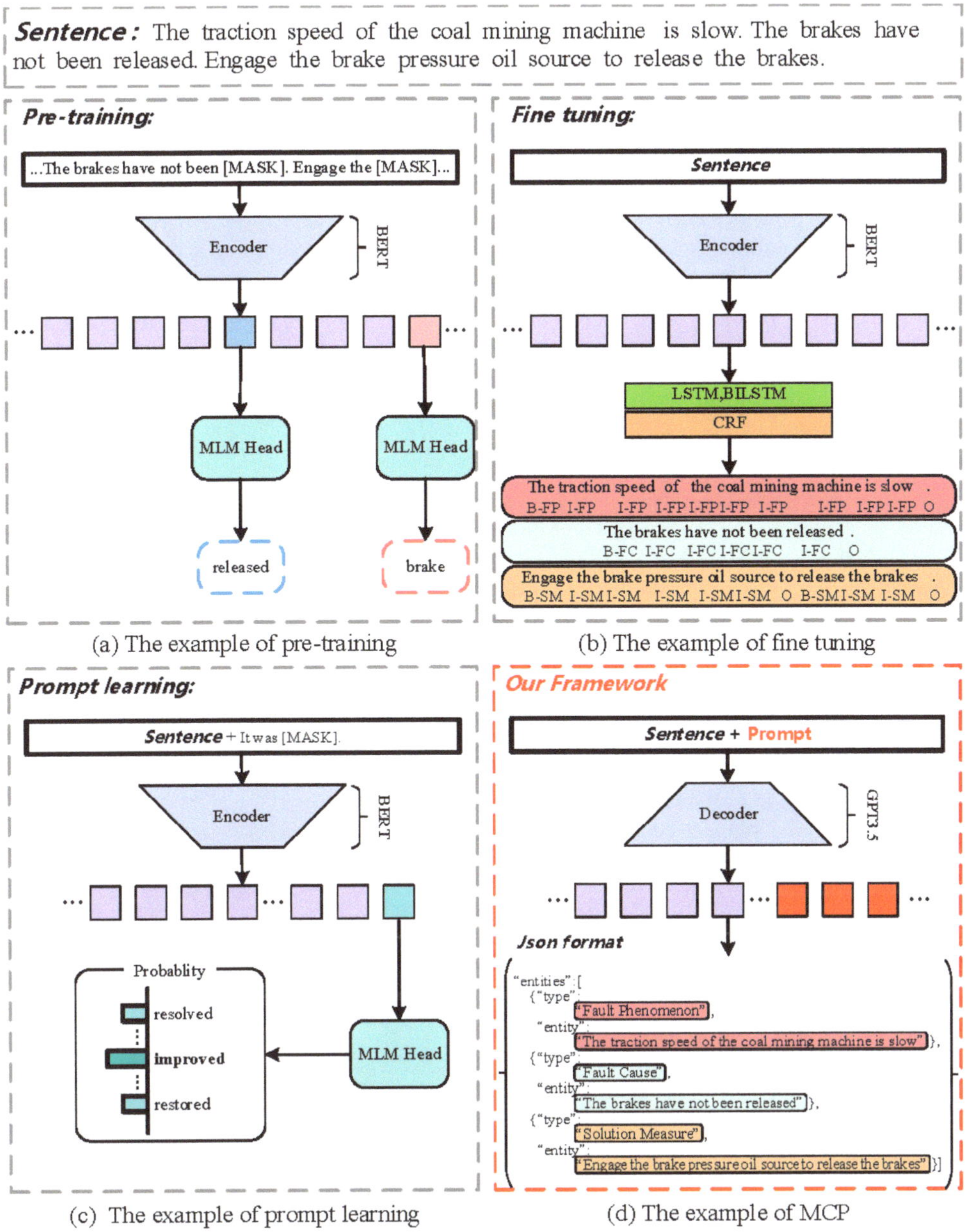

Figure 1. Comparison of the proposed and previous models.

Although information extraction methods based on prompt learning and LLMs have achieved remarkable results in various domains, their application to NER in shearer maintenance contexts faces two critical challenges. The first challenge lies in enabling LLMs to effectively process domain-specific technical terminology and complex logical relationships inherent in shearer maintenance documentation while mitigating potential comprehension errors and reasoning biases. The second challenge concerns designing specialized NER prompt templates that synergistically integrate domain knowledge with the linguistic processing capabilities of LLMs, particularly given the inherent limitations of generic prompt frameworks in handling contextually dense industrial corpora.

To address the above issues, this paper proposes a multi-level clue-guided prompt learning (MCP) method for NER in the shearer maintenance domain, as illustrated in Figure 1d. This method inputs the information to be identified into the MCP model, where the prompt learning layer encapsulates the information and multiple sub-prompts into structured prompts according to a predefined format. The connection of the structured prompts and the information to be identified is then fed into the LLM layer, which processes it through multiple stacked decoder layers for deep representation learning. The answer layer uses a parser to map the output of the LLM layer into a structured label space, yielding the recognition results for named entities in the shearer maintenance domain. The model extracts and learns trigger words related to entity recognition from the prompts to obtain context-aware prompt tokens. This enables the model to make accurate predictions without relying on fine-tuning with labeled data, while also reducing hallucination phenomena in the LLMs.

Overall, our significant contributions are as follows.

1. To address the gap between pretraining and fine-tuning tasks in existing NER methods for the shearer maintenance domain, as well as the high dependency on labeled data, this study introduces LLMs and prompt learning. In the shearer maintenance NER task, the proposed method, based on LLMs and prompt learning, demonstrates superior performance.
2. We design multi-level entity extraction sub-prompts that utilize contextual information to form effective prompt clues. Restriction rules are applied to prevent the model from making data extraction errors and exhibiting hallucination phenomena. Furthermore, ontological concepts specific to the shearer maintenance domain are introduced to inject domain knowledge into the prompts, enhancing the method's domain adaptability.
3. The proposed method is evaluated on a self-constructed shearer maintenance dataset and compared with mainstream baseline methods. Experimental results indicate that the proposed method outperforms the baselines in terms of precision, recall, and F1 scores, demonstrating the effectiveness of MCP. Additionally, ablation experiments, conducted by removing layer-by-layer prompt clues, further validate the role and effectiveness of each sub-prompt clue.

2. Related Work

2.1. NER in Coal Mining Field

In the domain of NER for coal mining, redeploying the BERT model through fine-tuning is currently the mainstream approach. Cao et al. [13] defined entities, relationships, and attributes of coal mining equipment knowledge to construct a schema-level ontology, and used the lattice LSTM model and event joint extraction model for knowledge extraction. Li et al. [14] employed the BiLSTM-CRF model to accomplish NER tasks in the domain of coal mining electromechanical equipment accidents. Liu et al. [15] utilized the lattice LSTM model for entity recognition in coal mine safety knowledge. Cao et al. [16] applied the BERT-BiLSTM-CRF model to NER in coal mining equipment maintenance. Xu et al. [17] employed the BERT-BiLSTM-CRF model for knowledge entity recognition in the field of coal mine construction safety. Dang et al. [18] focused on mine hoists and proposed a BERTw-BiLSTM-CRF-FL entity recognition method based on the BERT-BiLSTM-CRF model. Li et al. [19] improved the BERT model's training mechanism and developed CoalBERT, a pretrained dynamic model for semantic understanding of coal mine safety hazard texts. Qiu et al. [20] proposed an NER method for electromechanical mining equipment, AMEQA, which preliminarily determines long-sequence nested entity boundaries using a machine reading comprehension framework and employs a BiLSTM neural network with an integrated attention mechanism to explore contextual associations between entities.

2.2. NER Based on Prompt Learning

LLMs are pretrained on massive text datasets, endowing them with superior generalization and semantic understanding capabilities. Compared to traditional NER models, methods based on LLMs can better capture contextual information, identify named entities with higher accuracy, and exhibit stronger generalizability. Traditional fine-tuning methods are challenging to apply to LLMs with extensive parameters. Prompt learning was initially introduced to bridge the gap between upstream and downstream tasks [21–24]. However, researchers discovered that during inference with the GPT-3 model, introducing minimal textual prompts for specialized tasks can match the performance of state-of-the-art fine-tuned models without retraining. Furthermore, it can generate high-quality samples and demonstrate robust qualitative performance in dynamically defined tasks [25,26]. Consequently, NER methods based on prompt learning and LLMs have become a prominent research focus in the field of natural language processing.

Gilardi et al. [27] noted that ChatGPT outperforms crowdsourced annotators in multiple annotation tasks, including relevance, sentiment, topic recognition, and framing detection. Its zero-shot average accuracy exceeds that of crowdsourced workers by 25%, demonstrating higher consistency and achieving a 30% cost reduction. Pan et al. [28] explored ChatGPT's ability to extract complex entities and structures in zero-shot spoken language understanding and dialogue state tracking tasks using prompt-based approaches. Results from four benchmark experiments highlight the significant potential of large models in dialogue comprehension. González-Gallardo C. E. et al. [29] utilized ChatGPT to identify entities in historical documents. Studies [30,31] have shown that the quality of prompt templates greatly influences the performance of prompt-based models. Brown et al. [11] were the first to apply prompt-based learning for few-shot learning tasks in text classification. PPT [32] proposed a unified prompt template for various sentence classification tasks, including sentence-pair classification, multiple-choice classification, and single-sentence classification tasks. Han et al. [33] introduced rule-based prompt tuning and refinement (PTR) for multi-class text classification, leveraging logical rules to construct prompts with multiple sub-prompts. Liu et al. [34] proposed a prompt learning-based NER approach incorporating question answering, transforming the NER problem into a question-answering task. Wei et al. [35] proposed ChatIE, a multi-turn question-answering framework based on ChatGPT for zero-shot information extraction in general-domain texts. Van Zandvoort et al. [36] introduced sample prompting and pattern prompting strategies, showing that both significantly improved the performance of LLMs in medical dialogue summarization tasks based on ROUGE scores and human evaluations.

2.3. Differences from Other Methods

Unlike existing NER methods in the coal mining domain, our approach offers several distinct advantages. First, we introduce LLMs, which, compared to PLMs, can perform NER tasks in zero-shot or few-shot scenarios and demonstrate superior performance in long-tail entity recognition tasks. This capability makes LLMs particularly effective in shearer maintenance, a domain characterized by high annotation costs and predominantly lengthy entities. Secondly, LLMs often lack sufficient in-depth domain-specific knowledge, making it challenging to meet precise demands such as those in equipment maintenance. To address this, we propose the MCP model, which incorporates the ontology concepts specific to shearer maintenance, injecting domain knowledge into prompts. This approach reduces the need for domain-specific fine-tuning and mitigates issues like data extraction errors and hallucinations in the model.

3. Shearer Maintenance Field Ontology Modeling

This study constructed raw corpora for the shearer maintenance domain by collecting user manuals, maintenance guides, and repair case studies for various shearer models. Based on the characteristics of sample data and downstream task requirements, five categories of entity were defined: fault type, fault phenomenon, fault cause, solution measures, and fault location. Fault type refers to the classification of faults based on their nature, manifestation, and impact. Fault phenomenon describes the abnormal states or behaviors exhibited by equipment or systems when a fault occurs. Fault cause identifies the root causes that lead to equipment or system failures. Solution measures refer to the methods and steps taken to repair, adjust, or improve equipment or system failures. Fault location specifies the exact position or component within the equipment or system where the fault occurs. By collecting specialized terminology from the shearer maintenance domain, an initial domain-specific lexicon was constructed, as shown in Table 1.

Table 1. Domain dictionary.

Entity Type	Example
Fault category	Rocker arm operation failure
Fault phenomenon	The rocker arm cannot be raised or lowered
Failure cause	The solenoid valves are malfunctioning
Solution measures	Replace the damaged solenoid valves
Fault location	Rocker arm

The entities in the shearer maintenance domain do not exist in isolation, but are interconnected through semantic relationships. The semantic relationships in the shearer maintenance domain ontology are shown in Table 2.

Table 2. Semantic relationships.

Associated Concepts	Semantic Relationships	Concept Being Associated
Coal mining machine	has fault categories	Fault category
Fault category	manifests as	Fault phenomenon
Fault phenomenon	is attributed to	Failure cause
Failure cause	is resolved by	Solution measures
Fault category	occurs in	Fault location

To provide a more intuitive representation of the relationships between classes in the shearer maintenance domain, Figure 2 illustrates an example based on the aforementioned ontology concepts and relationship definitions, demonstrating the organized connections within shearer maintenance content.

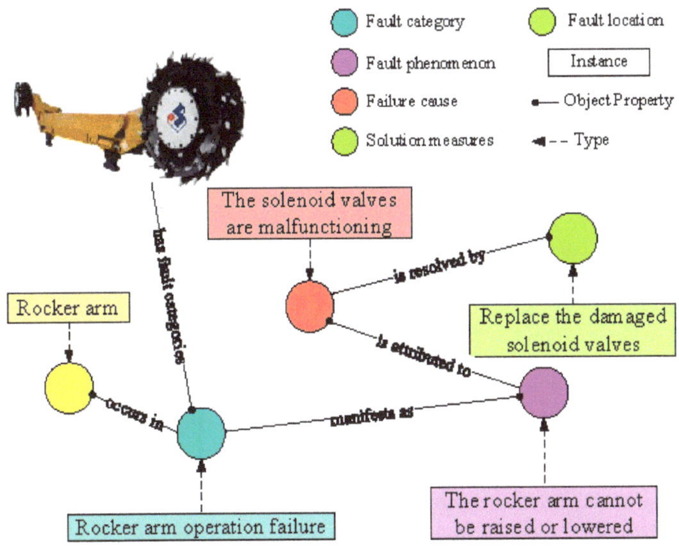

Figure 2. Example of an entity relationship diagram in the shearer maintenance domain.

4. Methodology

This section introduces the MCP model, which primarily consists of three components. The first component is the prompt learning layer, responsible for prompt design and construction. The second component is the decoder-only architecture-based LLM layer. The third component is the answer layer, which handles result formatting. The framework of the MCP model is illustrated in Figure 3.

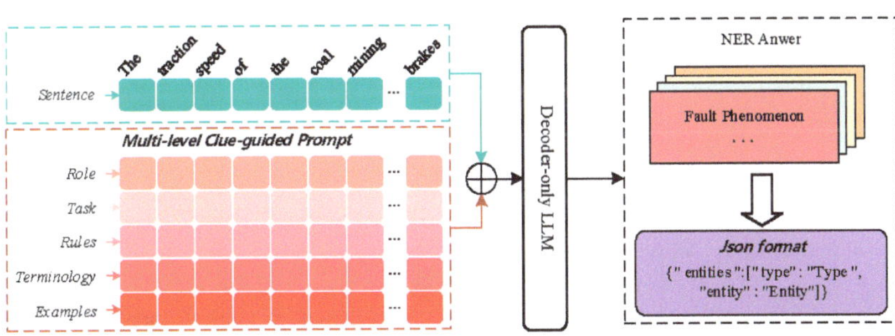

Figure 3. MCP model framework.

4.1. Task Definition

Given a sentence $S = \{x1, x2, \ldots, xn\}$ from the shearer maintenance domain, where n is the number of words and x_i represents the i-th word in S, the goal of the NER task is to predict the potential entities in sentence S, where $E = \{e_1, e_2, \ldots, e_m \mid S\}$ is a set of entities, each entity e_i is represented by its span (h_i, t_i) with h_i and t_i denoting the start and end positions of the entity in the sentence, and m is the number of entities.

4.2. Prompt Learning Layer

Nigh [37] collected a wealth of high-quality prompts and summaries related to the CRISPE framework, arguing that the core goal of prompt engineering is to maximize the utilization of LLMs' inherent generalization abilities and knowledge reserves. Prompts should be designed hierarchically based on complexity to ensure clarity and specificity, avoiding ambiguous or polysemous expressions, and collaboratively guiding the language model to produce accurate responses while minimizing hallucinated outputs. When the existing knowledge of general-purpose large models is insufficient to address tasks in specialized domains, additional background information can be introduced through inputting new knowledge to achieve higher-quality output results. Inspired by this, we referred to the CRISPE framework's design principles to improve prompts by encoding domain-specific knowledge into templates, thereby enriching textual context. A set of prompts $P = \{p_1, p_2, \ldots, p_n\}$ was constructed, and information was extracted by filling the slots within the prompts. Specifically, to tackle the challenges of technical terminology and contextual complexity, the prompt learning layer integrates domain knowledge through five sub-prompts.

$$\begin{aligned} Prompt &= \sum_{i=1}^{n} p_i \\ &= Role + Task + Rules + Terminology + Examples \end{aligned} \tag{1}$$

"Role" serves as role guidance, specifying the expected role of the LLMs to set its focus and improve its performance in specific tasks, thereby enhancing its understanding of domain-specific terminology and concepts. *"Task"* provides a task description, outlining the goals the user intends to achieve, which constitutes the specific content for the LLMs to perform the extraction task. *"Rules"* outlines detailed requirements, informing the model of the considerations and specific guidelines it must follow while performing the task. *"Terminology"* refers to specialized terms, injecting domain-specific knowledge into the model. Together with the *"Rules"* cues, it constrains the domain content and output entities, preventing the LLMs from excessive extraction that might introduce noise from other domains or cause hallucination issues. *"Examples"* refers to generated samples, providing examples of specific extraction tasks and demonstrating how to identify and extract embedded entities from a given text. These examples assist the model in understanding the specific task requirements and learning how to apply this knowledge in practical operations.

The structured prompts were dynamically generated using Python's Jinja2 template engine (Python 3.9), where *"Role"*, *"Task"*, *"Terminology"*, *"Rules"*, and *"Examples"* were injected via placeholders. *"Role"*: directs the model to act as a domain expert (e.g., "You are a shearer maintenance entity tagger"). *"Terminology"*: defines entity types (e.g., "Fault cause: root cause of equipment failure") to resolve ambiguity. *"Rules"*: constraints (e.g., "prefer longer spans") prevent fragmented extraction. *"Examples"*: demonstrates entity patterns (e.g., "Replace solenoid valves" → Solution). *"Task"*: specifies JSON output format to align with downstream applications. The specific template is shown in Table 3.

We encapsulated it to form structured prompt words, combining the original sentence S with the structured prompts as input for the LLMs. To formally describe the process of constructing structured prompts, we introduce a two-step formulation that clarifies the relationship between the original input text and the augmented prompt. The following equations define the core mechanism of our structured prompt generation method, emphasizing both the functional abstraction and its concrete implementation.

$$S_{prompt} = f_{prompt}(S) \tag{2}$$

$$f_{prompt}(S) = S \| Prompt \tag{3}$$

Explanation:

Equation (2) declares that the final prompt S_{prompt} is generated by applying the function f_{prompt} to the original input sentence S.

Equation (3) defines the internal logic of f_{prompt}, where || denotes the template-aware concatenation operator. This operation dynamically inserts S into a predefined structured prompt template.

Table 3. Example of internal elements of the prompt.

Sub-Prompt	Sub-Prompt Example
Role	You are a shearer maintenance domain entity tagger and need to complete the task of tagging entities for a given text fragment.
Task	This task requires you to identify five types of entities in the text: fault category, fault symptom, fault cause, solution, fault location, and format the output as a JSON object in the format of {"entities": ["type": "Type", "entity": "Entity"]}.
Rules	• Please do not output any superfluous words other than JSON. • In the case of ambiguity, it always tends to have a longer physical span. • Do not change the original field, output redundant punctuation marks. • According to the entity definition, only the five types of entities are identified, and when there is no type, the type does not need to be output.
Terminology	• Fault category: refers to the classification of faults according to the nature, performance, and impact of faults. • Fault phenomenon: refers to the abnormal state or behavior of the equipment or system when it fails. • Failure cause: guide to the root cause of equipment or system failure. • Solution measures: refers to the methods and steps taken to repair, adjust, or improve the equipment or system failure. • Fault location: refers to the specific location or component of the fault in the equipment or system.
Examples	• The traction speed of the coal mining machine is slow. The brakes have not been released. Engage the brake pressure oil source to release the brakes. • {"entities": [{"type": "Fault Phenomenon", "entity": "The traction speed of the coal mining machine is slow"}, {"type": "Fault Cause", "entity": "The brakes have not been released"}, {"type": "Solution Measure", "entity": "Engage the brake pressure oil source to release the brakes"}]}

4.3. LLM Layer

The model employed in this study is GPT-3.5-turbo, accessed through the OpenAI API. It represents a widely recognized decoder-only architecture within transformer-based LLMs. The parameters were configured as temperature = 0.1 and max_tokens = 1000 to balance the accuracy and diversity of the generated results. This model increases its depth and expressive capacity by stacking multiple identical decoder layers, and each layer's structure roughly matches that shown in Figure 4. The first layer receives token sequences encoded by the word embedding layer. Each layer consists of two sub-layers: the first is a masked multi-head self-attention (MHA) layer, and the second is a feed-forward network (FFN) layer. Each sub-layer employs residual connections, followed by layer normalization [38] immediately after the addition operation of the residual connections [39,40]. Subsequent layers take the hidden states output by the previous layer as input, process them through the MHA and FFN layers, and pass the results to the next layer. The hidden states output by the final layer are fed into a linear layer called the language modeling head. The resulting logits are then converted into a probability distribution over target vocabulary using a softmax operation.

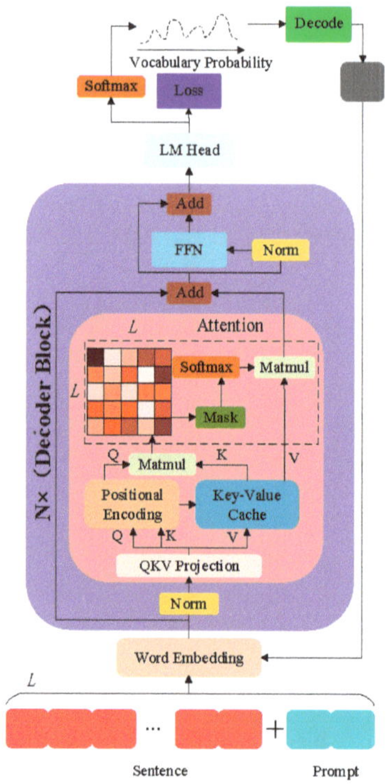

Figure 4. Transformer-based Decoder-only LLMs.

The attention mechanism is a core design component of LLMs based on the decoder-only architecture, implemented in the masked MHA layer [40]. As shown in Figure 4, the original sentence S concatenated with the structured prompt serves as the input to the LLM, with a total length of L. At this stage, the token sequence of word embeddings is represented as $X \in \mathbb{R}^{L \times d_{in}}$, generating three embedding matrices: *query* $Q \in \mathbb{R}^{L \times d_q}$, *key* $K \in \mathbb{R}^{L \times d_k}$, and *value* $V \in \mathbb{R}^{L \times d_v}$ through the linear projection layer. The model illustrated in the diagram employs relative positional encoding (e.g., rotary position embedding, RoPE [41]). RoPE primarily applies to Q and K to influence attention computation. The V does not require positional encoding, as it only participates in the weighted sum and does not directly impact attention weight calculation. RoPE modifies Q and K via rotational transformation, allowing attention weights to capture relative positional information. Since V is only used for weighted summation, it does not require additional positional information.

$$Q, K, V := \text{split}(X \times W_{q,k,v}),\ W_{q,k,v} \in \mathbb{R}^{d_{in} \times (d_q + d_k + d_v)} \tag{4}$$

Then, for the attention kernel operation:

$$P := Q \times K^T,\ A := \text{softmax}\left[\frac{\mathrm{P}}{\sqrt{\mathrm{d_k}}} \odot M\right],\ O := (A \times V) \times W_O, W_O \in \mathbb{R}^{d_v \times d_o} \tag{5}$$

That is, each entry of the unnormalized relevance matrix $P \in \mathbb{R}^{L \times L}$ measures the correlation between the corresponding token pairs. The normalized attention score matrix $A \in \mathbb{R}^{L \times L}$ is obtained by scaling with the factor $\sqrt{d_k}$, combined with the element-wise masking operation $M \in \mathbb{R}^{L \times L}$ and the softmax calculation along the row direction. Finally, the output hidden state $O \in \mathbb{R}^{L \times d_o}$ is generated by a weighted sum of V using the attention weights of each row of A, usually followed by an additional linear transformation.

It is important to note that the embedding dimensions of the Q, K, V, O do not necessarily have to be the same. In this paper, although subscripts are used to distinguish them for generality, we assume that $d = d_q = d_k = d_v = d_o$ is used to capture diverse relationships. The model typically uses multi-head attention rather than single-head attention to achieve this, by splitting learnable parameters such as $W_{q,k,v} \in \mathbb{R}^{d_{in} \times (3 \times d)}$ into $W^{mh}_{q,k,v} \in \mathbb{R}^{d_{in} \times (3 \times H \times d_{head})}$, where H represents the number of heads. Attention is executed in parallel across different sets of Q_h, K_h, and V_h for each attention head. Similar to the embedding dimension, the number of heads for Q, K, and V may vary across different LLMs, but in this paper, we assume they are the same.

4.4. Answer Layer

Prompt learning helps the model establish a mapping between the input X and the task label space T, transforming the generated predictions into the desired final answer format. Specifically, the LLM calculates the probability distribution after processing the input with the prompt template. From a Bayesian perspective, the output probability is:

$$p(y|x + prompt) = \int_{concept} p(y|x + prompt, concept)\, p(concept|x + prompt)\, d(concept) \quad (6)$$

Here, y represents the output, x represents the extracted input text, and *concept* denotes the underlying logic containing statistical information about the text. The inference process of the result can be viewed as a Bayesian inference of a shared concept from the input. If the model can infer the shared concept, it can make accurate predictions on test samples. This can be interpreted as the conditional probability of inferring the shared concept from the input. Mathematically, x + *prompt* provides evidence to the model, enhancing the posterior distribution of the concept. The output of the LLM is processed through a designed parser R, mapping it to the structured label space O, thereby generating structured shearer maintenance domain entity data. The parser begins by using regular expressions to extract JSON-formatted entity fragments from raw text. It then converts these fragments into Python dictionaries using the json library (via json.loads()) in Python 3.9. Finally, the application of custom validation rules filters out invalid entries, ensuring that only clean, usable data remain for analysis.

5. Experiments

We conducted a series of experiments to validate the effectiveness of the MCP model. We utilized a dataset from the shearer maintenance domain to evaluate the performance of our model on the NER task. The experimental results demonstrate that the proposed method outperforms baseline models.

5.1. Dataset Construction

Given the lack of sufficient annotated corpora in the shearer maintenance domain, this study independently constructed a dataset for this domain by collecting manuals, maintenance guides, and repair case studies for various shearer models. Content that did not meet quality and relevance criteria was removed, while the original format and expressions of domain-specific materials were preserved. Data in the shearer maintenance domain are characterized by long texts, lack of subjects, frequent use of technical terms, and irregular structures, which add complexity to entity recognition tasks in this field. Data cleaning was conducted by removing formatting issues and garbled characters. Subsequently, format standardization was performed to convert documents of various formats into a unified text format (e.g., plain text or HTML). The text was segmented, non-ASCII characters were removed, and encoding was standardized. Noise reduction was then applied to the text, which included removing spaces, special symbols, and meaningless short sentences, and using spell-check and correction tools to rectify typos. Finally, missing knowledge was supplemented or relevant additional information was provided to refine the meaning of textual expressions. The dataset for the shearer maintenance domain was constructed using the BIO tagging method. The dataset contained a total of 1092 sentences and 2158 entities, with 80% of the dataset randomly selected for baseline model training and the remaining 20% used for model testing. The distribution of entity types is shown in Figure 5.

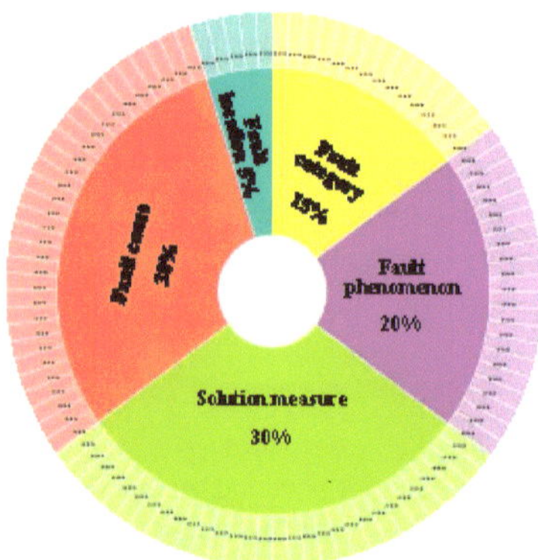

Figure 5. Distribution of entity types by proportion.

5.2. Baselines

This section compares the proposed method with the mainstream models for NER in the shearer maintenance domain: BERT-BiLSTM-CRF and BiLSTM-CRF.

BiLSTM-CRF [42]. This model is a classic NER model that leverages BiLSTM to capture contextual features and predicts entity labels for each character. The CRF algorithm is used to propagate the probability matrix and control the output of sequence labeling.

BERT-BiLSTM-CRF [43]. This model embeds contextual information for each character using the BERT layer, extracts text features with the BiLSTM model, and employs the CRF layer to constrain the relationships between labels. It is currently a mainstream approach for addressing sequence labeling tasks.

5.3. Evaluation Metrics

We used three standard evaluation metrics to assess the models in the experiments: precision (Prec.), recall (Rec.), and F1 score (F1). Precision indicates the proportion of correctly identified entities among all identified entities, while recall represents the proportion of correctly identified entities among all ground truth entities. The F1 score is calculated as the harmonic mean of precision and recall. The calculations for P, R, and F1 score are shown in the formulas, where TP represents the number of true positives, FP represents the number of false positives, and FN represents the number of false negatives.

$$P = \frac{TP}{TP + FP} \times 100\% \tag{7}$$

$$R = \frac{TP}{TP + FN} \times 100\% \tag{8}$$

$$F1 = \frac{2PR}{P + R} \times 100\% \tag{9}$$

5.4. Comparison Experiments

Table 4 shows the test results of the proposed method and two baseline models. It can be observed from Table 4 that the proposed method outperforms traditional fine-tuning-based methods on all metrics, achieving a 4.85% improvement in the F1 score compared to the BiLSTM-CRF model and a 1.19% improvement compared to the BERT-BiLSTM-CRF model. Compared to fine-tuning-based methods, such as BERT-BiLSTM-CRF, which employ carefully designed downstream classifiers and incorporate external knowledge through pretraining to enhance model performance, their performance still lags behind the prompt-based learning approach. The primary reason for this disparity is the gap between upstream and downstream task objectives in the fine-tuning paradigm. In contrast, prompt-based learning methods can more effectively leverage the knowledge and capabilities of language models without requiring additional training. This is particularly valuable in domains such as shearer maintenance, where annotation costs are high and data is scarce.

Table 4. Results of entity recognition experiment.

Method	P	R	F1
BiLSTM-CRF	82.97%	81.85%	82.41%
BERT-BiLSTM-CRF	86.75%	85.40%	86.07%
MCP	87.45%	87.08%	87.26%

5.5. Ablation Study

We conducted five ablation experiments by sequentially removing the *Role*, *Task*, *Terminology*, *Rules*, and *Examples* layers. These experiments aimed to validate that the shearer maintenance entity recognition method, guided by multi-level collaborative prompts, improves model accuracy and mitigates output instability.

To evaluate the impact of the five sub-prompts on model performance, we conducted ablation studies. In these, each sub-prompt was removed individually and the model's performance was compared. The results of the ablation experiments are shown in Figure 6. As shown in Figure 6, (i) when the *Role* sub-prompt is removed, the F1 score for entity extraction decreases by approximately 13% compared to our proposed method. This is because assigning a specific role to the model enables it to better understand the context of the problem and user expectations, leading to more accurate and consistent responses. The *Role* sub-prompt helps the model focus on domain-specific information and knowledge, enabling it to provide more specialized answers. Removing the *Role* sub-prompt introduces noise, resulting in a decline in the F1 score for entity extraction. (ii) When the *Task* sub-prompt is removed, the F1 score decreases by approximately 16%. This is because the Task sub-prompt helps the model clearly understand the task objectives and expected outcomes, thereby reducing misunderstandings and errors. (iii) When the *Terminology* sub-prompt is removed, the precision and F1 score for entity extraction decrease by 18.44% and 10.04%, respectively. This is because the *Terminology* sub-prompt provides the model with relevant definitions of terms in the shearer maintenance domain. After removing the *Terminology* sub-prompt, the model lacks domain-specific knowledge and constraints, resulting in a significant decline in precision. (iv) When the *Rules* sub-prompt is removed, the F1 score for entity extraction decreases by approximately 9.0%. This is because the *Rules* sub-prompt emphasizes output consistency and ensures data completeness, enabling the model to focus on specific requirements. (v) When the *Examples* sub-prompt is removed, the F1 score for entity extraction decreases by 12%. This is because *Examples* serve as part of the training data, helping the model better learn the characteristics and patterns of the specific task. By providing clear examples, the model can more accurately understand the task details and boundary conditions, reducing errors when handling complex or ambiguous tasks.

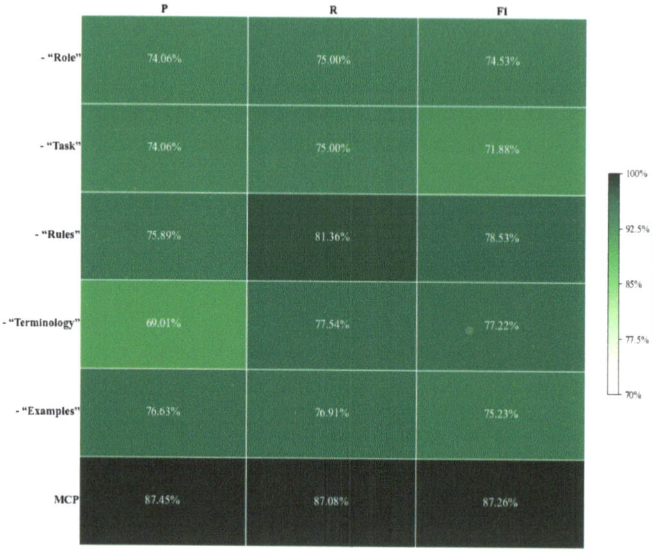

Figure 6. Results of ablation experiments.

The experimental results above indicate that the five designed sub-prompts all enhance model performance to varying degrees, demonstrating the effectiveness and necessity of each sub-prompt.

6. Discussion

6.1. Addressing Technical Term Challenges

The shearer maintenance domain contains terms (e.g., "rocker arm operation failure") that are rare in general corpora. Traditional LLMs like GPT-3 often misinterpret such terms due to training data bias [11]. By encoding domain terminology into prompts (Table 3), MCP aligns LLMs' latent knowledge with expert-defined concepts similarly to medical NER methods that inject clinical lexicons [36].

6.2. Domain-Adaptive Prompt Design

Generic prompts (e.g., "extract all entities") fail in industrial contexts with nested entities (e.g., "fault location → rocker arm"). MCP's multi-level prompts decompose extraction into role-specific steps (e.g., "identify fault causes first"), mimicking human expert workflows. This echoes studies on task decomposition in low-resource IE [34], but extends it through ontology-guided constraints.

7. Conclusions

While general-domain NER algorithms have advanced significantly, their adaptation to specialized industrial applications such as shearer maintenance remains limited, with conventional fine-tuning approaches dominating existing solutions. This study addresses this gap by proposing a multi-level clue-guided prompt learning (MCP) framework that harnesses the few-shot learning capabilities of LLMs to achieve resource-efficient named entity recognition in shearer maintenance scenarios. Our experimental evaluation demonstrates the superiority of MCP over mainstream baselines, yielding F1-score improvements of 4.85% (vs. BiLSTM-CRF) and 1.19% (vs. BERT-BiLSTM-CRF), while eliminating dependency on large annotated datasets and exhibiting enhanced generalization potential.

Our study has several limitations that warrant further investigation. First, the current MCP framework's performance is contingent upon manually engineered prompt templates, which introduces subjectivity and may compromise robustness across diverse maintenance contexts. Second, while achieving state-of-the-art results on shearer maintenance datasets, the method's generalizability to other industrial domains (e.g., hydraulic systems or mining conveyors) remains unverified.

Future research should focus on three strategic directions to address these challenges. Automated prompt optimization strategies, potentially leveraging reinforcement learning or gradient-based template search, could minimize human intervention in template design. Cross-domain validation studies are needed to systematically quantify the framework's adaptability to broader industrial NLP tasks, including fault diagnosis in interconnected mechanical systems. Finally, hybrid architectures integrating LLM-based few-shot learning with model compression techniques (e.g., knowledge distillation) or traditional supervised models may offer viable pathways to balance accuracy, computational efficiency, and extreme low-resource adaptability.

Author Contributions: Methodology, X.C. and L.S.; Software, X.W. and X.Y.; Validation, X.Z.; Data curation, L.S. and Y.D.; Writing—original draft, X.C. and L.S.; Writing—review and editing, X.C. and L.S. All authors have read and agreed to the published version of the manuscript.

Funding: This work was supported by the Natural Science Foundation of China (52274158 and 51834006).

Institutional Review Board Statement: Not applicable.

Informed Consent Statement: Not applicable.

Data Availability Statement: The data and materials used in this research are available upon reasonable request to the corresponding author. The data are not publicly available due to privacy.

Conflicts of Interest: The authors declare no conflicts of interest.

References

1. Sun, P.; Yang, X.; Zhao, X.; Wang, Z. An overview of named entity recognition. In Proceedings of the 2018 International Conference on Asian Language Processing (IALP), Bandung, Indonesia, 15–17 November 2018; pp. 273–278.
2. Wen, Y.; Fan, C.; Chen, G.; Chen, X.; Chen, M. A survey on named entity recognition. In Proceedings of the 8th International Conference on Communications, Signal Processing, and Systems, Changbaishan, China, 4–5 July 2020; pp. 1803–1810.
3. Zhou, G.; Su, J. Named entity recognition using an HMM-based chunk tagger. In Proceedings of the 40th Annual Meeting of the Association for Computational Linguistics, Philadelphia, PA, USA, 6–12 July 2002; pp. 473–480.
4. Tsai, R.T.-H.; Hung, H.-C.; Sung, C.-L.; Dai, H.-J.; Hsu, W.-L. On closed task of Chinese word segmentation: An improved CRF model coupled with character clustering and automatically generated template matching. In Proceedings of the Fifth SIGHAN Workshop on Chinese Language Processing, Sydney, Australia, 22–23 July 2006; pp. 134–137.
5. Fresko, M.; Rosenfeld, B.; Feldman, R. A hybrid approach to NER by MEMM and manual rules. In Proceedings of the 14th ACM International Conference on Information and Knowledge Management, Bremen, Germany, 31 October–5 November 2005; pp. 361–362.
6. Isozaki, H.; Kazawa, H. Efficient support vector classifiers for named entity recognition. In Proceedings of the COLING 2002: The 19th International Conference on Computational Linguistics, Taipei, Taiwan, 24 August–1 September 2002.
7. Huang, Z.; Xu, W.; Yu, K. Bidirectional LSTM-CRF models for sequence tagging. *arXiv* **2015**, arXiv:1508.01991.
8. Devlin, J. Bert: Pre-training of deep bidirectional transformers for language understanding. *arXiv* **2018**, arXiv:1810.04805.
9. Gao, T.; Fisch, A.; Chen, D. Making pre-trained language models better few-shot learners. *arXiv* **2020**, arXiv:2012.15723.
10. Petroni, F.; Rocktäschel, T.; Lewis, P.; Bakhtin, A.; Wu, Y.; Miller, A.H.; Riedel, S. Language models as knowledge bases? *arXiv* **2019**, arXiv:1909.01066.
11. Brown, T.; Mann, B.; Ryder, N.; Subbiah, M.; Kaplan, J.D.; Dhariwal, P.; Neelakantan, A.; Shyam, P.; Sastry, G.; Askell, A. Language models are few-shot learners. *Adv. Neural Inf. Process. Syst.* **2020**, *33*, 1877–1901.
12. OpenAI. Chatgpt: Optimizing Language Models for Dialogue. 2022. Available online: https://chatgpt.r4wand.eu.org/ (accessed on 12 October 2024).

13. Cao, X.; Zhang, M.; Lei, Z.; Duan, X.; Chen, R. Construction and application of knowledge graph for coal mine equiprent maintenance. *Ind. Mine Autom.* **2021**, *42*, 41–45.
14. Li, Z.; Zhou, B.; Li, W.; Zhou, Y.; Feng, Z.; Zhao, H. Construction and application of mine electromechanical equipment accident knowledge graph. *Ind. Mine Autom.* **2022**, *48*, 109–112.
15. Liu, P.; Ye, S.; Shu, Y.; Lu, X.; Liu, M. Coalmine Safety: Knowledge Graph Construction and Its QA Approach. *J. Chin. Inf. Process.* **2020**, *34*, 49–59.
16. Cao, X.; Wu, K.; Zhang, M.; Duan, Y.; Li, P. Coal mine equipment maintenance knowledge named entity recognition model based on BERT. *Mach. Tool Hydraul.* **2023**, *51*, 103–108.
17. Xu, N.; Liang, Y.; Guo, C.; Meng, B.; Zhou, X.; Hu, Y.; Zhang, B. Entity recognition in the field of coal mine construction safety based on a pre-training language model. *Eng. Constr. Archit. Manag.* **2023**. [CrossRef]
18. Dang, X.; Liu, J.; Dong, X.; Zhu, Z.; Li, F. Named entity recognition of mechanical equipment failure for imbalanced data. *Comput. Eng.* **2024**, *50*, 104–112.
19. Li, Z.; Liu, F.; Zhao, J.; Qi, H.; Li, J. Construction of pre-training language model for coal mine safety hidden danger texts. *Min. Saf. Environ. Prot.* **2024**. *Ahead of Print*.
20. Qiu, Y.; Xing, H.; Yu, Z.; Zhang, W. Named entity recognition for mine electromechanical equipment monitoring text. *Comput. Eng. Appl.* **2024**, *60*, 129–138.
21. Chen, X.; Zhang, N.; Xie, X.; Deng, S.; Yao, Y.; Tan, C.; Huang, F.; Si, L.; Chen, H. Knowprompt: Knowledge-aware prompt-tuning with synergistic optimization for relation extraction. In Proceedings of the ACM Web Conference 2022, Lyon, France, 25–29 April 2022; pp. 2778–2788.
22. Chen, Y.; Liu, Y.; Dong, L.; Wang, S.; Zhu, C.; Zeng, M.; Zhang, Y. Adaprompt: Adaptive model training for prompt-based nlp. *arXiv* **2022**, arXiv:2202.04824.
23. Chia, Y.K.; Bing, L.; Poria, S.; Si, L. RelationPrompt: Leveraging prompts to generate synthetic data for zero-shot relation triplet extraction. *arXiv* **2022**, arXiv:2203.09101.
24. Lu, Y.; Liu, Q.; Dai, D.; Xiao, X.; Lin, H.; Han, X.; Sun, L.; Wu, H. Unified structure generation for universal information extraction. *arXiv* **2022**, arXiv:2203.12277.
25. Schick, T.; Schütze, H. Exploiting cloze questions for few shot text classification and natural language inference. *arXiv* **2020**, arXiv:2001.07676.
26. Schick, T.; Schütze, H. It's not just size that matters: Small language models are also few-shot learners. *arXiv* **2020**, arXiv:2009.07118.
27. Gilardi, F.; Alizadeh, M.; Kubli, M. ChatGPT outperforms crowd workers for text-annotation tasks. *Proc. Natl. Acad. Sci. USA* **2023**, *120*, e2305016120. [CrossRef]
28. Pan, W.; Chen, Q.; Xu, X.; Che, W.; Qin, L. A preliminary evaluation of chatgpt for zero-shot dialogue understanding. *arXiv* **2023**, arXiv:2304.04256.
29. González-Gallardo, C.-E.; Boros, E.; Girdhar, N.; Hamdi, A.; Moreno, J.G.; Doucet, A. Yes but.. can chatgpt identify entities in historical documents? In Proceedings of the 2023 ACM/IEEE Joint Conference on Digital Libraries (JCDL), Santa Fe, NM, USA, 26–30 June 2023; pp. 184–189.
30. Lu, Y.; Bartolo, M.; Moore, A.; Riedel, S.; Stenetorp, P. Fantastically ordered prompts and where to find them: Overcoming few-shot prompt order sensitivity. *arXiv* **2021**, arXiv:2104.08786.
31. Webson, A.; Pavlick, E. Do prompt-based models really understand the meaning of their prompts? *arXiv* **2021**, arXiv:2109.01247.
32. Gu, Y.; Han, X.; Liu, Z.; Huang, M. Ppt: Pre-trained prompt tuning for few-shot learning. *arXiv* **2021**, arXiv:2109.04332.
33. Han, X.; Zhao, W.; Ding, N.; Liu, Z.; Sun, M. Ptr: Prompt tuning with rules for text classification. *AI Open* **2022**, *3*, 182–192. [CrossRef]
34. Liu, A.T.; Xiao, W.; Zhu, H.; Zhang, D.; Li, S.-W.; Arnold, A. Qaner: Prompting question answering models for few-shot named entity recognition. *arXiv* **2022**, arXiv:2203.01543.
35. Wei, X.; Cui, X.; Cheng, N.; Wang, X.; Zhang, X.; Huang, S.; Xie, P.; Xu, J.; Chen, Y.; Zhang, M. Chatie: Zero-shot information extraction via chatting with chatgpt. *arXiv* **2023**, arXiv:2302.10205.
36. van Zandvoort, D.; Wiersema, L.; Huibers, T.; van Dulmen, S.; Brinkkemper, S. Enhancing Summarization Performance through Transformer-Based Prompt Engineering in Automated Medical Reporting. *arXiv* **2023**, arXiv:2311.13274.
37. Shieh, J. Best practices for prompt engineering with OpenAI API. *OpenAI*. February 2023. Available online: https://help.openai.com/en/articles/6654000-best-practices-for-prompt-engineering-with-openai-api (accessed on 24 October 2024).
38. Lei Ba, J.; Kiros, J.R.; Hinton, G.E. Layer normalization. *arXiv* **2016**, arXiv:1607.06450.
39. He, K.; Zhang, X.; Ren, S.; Sun, J. Deep residual learning for image recognition. In Proceedings of the 2016 IEEE Conference on Computer Vision and Pattern Recognition (CVPR), Las Vegas, NV, USA, 27–30 June 2016; pp. 770–778.
40. Vaswani, A. Attention is all you need. In Proceedings of the Advances in Neural Information Processing Systems 30: Annual Conference on Neural Information Processing Systems 2017, Long Beach, CA, USA, 4–9 December 2017.

41. Su, J.; Ahmed, M.; Lu, Y.; Pan, S.; Bo, W.; Liu, Y. Roformer: Enhanced transformer with rotary position embedding. *Neurocomputing* **2024**, *568*, 127063. [CrossRef]
42. Lample, G. Neural architectures for named entity recognition. *arXiv* **2016**, arXiv:1603.01360.
43. Jia, C.; Shi, Y.; Yang, Q.; Zhang, Y. Entity enhanced BERT pre-training for Chinese NER. In Proceedings of the 2020 Conference on Empirical Methods in Natural Language Processing (EMNLP), Online, 16–20 November 2020; pp. 6384–6396.

applied sciences

MDPI

Article

Multi-Task Supervised Alignment Pre-Training for Few-Shot Multimodal Sentiment Analysis

Junyang Yang, Jiuxin Cao * and Chengge Duan

School of Cyber Science and Engineering, Southeast University, Nanjing 211189, China; junyang_yang@seu.edu.cn (J.Y.); secrecyge@126.com (C.D.)
* Correspondence: jx.cao@seu.edu.cn

Abstract: Few-shot multimodal sentiment analysis (FMSA) has garnered substantial attention due to the proliferation of multimedia applications, especially given the frequent difficulty in obtaining large quantities of training samples. Previous works have directly incorporated vision modality into the pre-trained language model (PLM) and then leveraged prompt learning, showing effectiveness in few-shot scenarios. However, these methods encounter challenges in aligning the high-level semantics of different modalities due to their inherent heterogeneity, which impacts the performance of sentiment analysis. In this paper, we propose a novel framework called Multi-task Supervised Alignment Pre-training (MSAP) to enhance modality alignment and consequently improve the performance of multimodal sentiment analysis. Our approach uses a multi-task training method—incorporating image classification, image style recognition, and image captioning—to extract modal-shared information and stronger semantics to improve visual representation. We employ task-specific prompts to unify these diverse objectives into a single Masked Language Model (MLM), which serves as the foundation for our Multi-task Supervised Alignment Pre-training (MSAP) framework to enhance the alignment of visual and textual modalities. Extensive experiments on three datasets demonstrate that our method achieves a new state-of-the-art for the FMSA task.

Keywords: few-shot multimodal sentiment analysis; multi-task learning; supervised pre-training

1. Introduction

The proliferation of multimedia applications has propelled multimodal sentiment analysis (MSA) on image and text to the forefront of research endeavors. Whether it is for dialogue systems or human–computer interaction, or even more application scenarios, MSA has a wide range of applicability [1]. However, early MSA research was limited by large-scale manual annotation work, requiring careful labeling for a large number of sentiment tags, which undoubtedly consumed a lot of cost and time [2,3]. FMSA, on the other hand, can achieve accurate sentiment analysis by efficiently leveraging a small number of samples in resource-limited scenarios. For this reason, exploring few-shot multimodal sentiment analysis (FMSA) in resource-poor scenarios in the real world is gradually showing its feasibility.

Prompt learning has demonstrated effectiveness for few-shot learning in natural language processing (NLP) by reconfiguring the downstream task into the pre-training task so that the pre-trained language model (PLM) requires only minimal training data to adapt to the downstream task [4]. PVLM [5] leverages prompt learning in multimodal few-shot scenarios and directly incorporates vision modality into the PLM. However, the challenges brought by heterogeneity cannot be ignored. This fundamental difference between visual and language modalities impedes PLM from directly parsing and comprehending visual information. Faced with this situation, a research direction of few-shot multimodal sentiment analysis (FMSA) is how to effectively align visual and language modalities to obtain consistent visual-language representations.

Inspired by recent studies on the Multimodal Large Language Models (MLLM) [6–8], we recognize that alignment pre-training is a potent way to tackle this issue. Implementing this strategy would aid in enhancing the PLM's flexible adaptability to the visual modality. UP-MPF [9], the previous state-of-the-art method for FMSA, leverages alignment pre-training to address the challenge of modality alignment in FMSA. To achieve alignment pre-training, UP-MPF freezes the PLM and feeds a substantial quantity of images along with text prompts into the PLM to predict the direction of image rotation in textual format, thereby enabling the model to learn the relationship between low-level visual features and language features, effectively enhancing the alignment between these modalities. However, UP-MPF exposes two main issues in practice: (1) The task of predicting the direction of image rotation, a well-known self-supervised pretext task in computer vision, primarily models low-level visual features and lacks the complex semantics that include human emotions [10]. These supervision signals, which are absent in this task, are crucial for aligning the learned visual features with strongly semantic language features. (2) UP-MPF only uses a single task during the pre-training stage. Focusing solely on one pre-training task can limit a model's visual understanding by exposing it to only a narrow range of contexts and patterns, making it less adept at comprehensively analyzing diverse visual information. This could hinder the model's effective learning of shared information between visual and language modalities, thereby presenting an obstacle to effective modality alignment.

In summary, the enhancement of few-shot multimodal sentiment analysis performance hinges on the optimization of the alignment pre-training stage. This optimization aims to bolster the model's visual grounding capabilities and foster a more comprehensive multimodal understanding, thereby empowering downstream sentiment analysis tasks. By doing so, the pre-trained model can acquire robust sentiment analysis proficiency through a limited number of training samples in data-scarce scenarios. Current alignment pre-training approaches exhibit deficiencies in rich semantic information and modal sharing, which are crucial elements for few-shot multimodal sentiment analysis tasks. Consequently, our research focuses on addressing these limitations to improve model performance in few-shot multimodal sentiment analysis contexts. By refining the alignment pre-training process, we aim to bridge the gap between visual and textual modalities, enabling more effective knowledge transfer and generalization in low-resource settings. This approach has the potential to significantly advance the field of multimodal sentiment analysis, particularly in scenarios where labeled data are scarce.

To address the above issues, we propose Multi-task Supervised Alignment Pre-training (MSAP) for the FMSA task. For issues with missing rich semantic information, since more and more large-scale labeled datasets designed for tasks such as image classification or image captioning have become easily accessible [11], our method combines these datasets for supervised pre-training [12], which provide explicit semantic signals like image categories, styles, or descriptions. This makes it possible to align visual and language modalities by incorporating stronger semantic information into the visual representation. Regarding the issue of missing modal shared information, we recognize that relying on a single pre-training task during the alignment stage may lead to potential overfitting and hinder the model's ability to learn generalized multimodal representations. This limitation could result in the model becoming overly biased, thereby compromising its effectiveness in capturing shared information across different modalities. To address this issue, we employ multi-task learning [13], a paradigm that enables the model to synthesize knowledge acquired from diverse tasks, thereby mitigating the risk of overfitting to any single task [14]. Moreover, multi-task learning facilitates the extraction of shared information across modalities, enhancing the model's ability to leverage cross-modal relationships. This approach not only promotes a more balanced and generalizable learning process but also fosters the development of robust representations that capture the intricate interplay between different modalities.

Specifically, we use a vision encoder to extract visual features and use a simple single-layer MLP as the alignment layer to align visual features with language features. Then, we feed the aligned visual tokens and language tokens into the pre-trained language model. During the Multi-task Supervised Alignment Pre-training (MSAP) stage, we first select three large image datasets, each annotated for different tasks. Combining these datasets, we created a massive comprehensive dataset containing approximately 900,000 images. As shown in Figure 1, our strategy is to fix the pre-trained language model (PLM) and update the weights of the visual encoder and alignment layer. Through the transformation of diverse visual tasks—encompassing image classification, image style recognition, and image captioning—into a unified Masked Language Modeling (MLM) paradigm, we achieve a synergistic integration of supervised pre-training and multi-task learning methodologies. Doing so not only improves the overall learning effect through the complementarity of each task but also allows the visual features to obtain richer semantic information. In the prompt fine-tuning stage, we fix the visual encoder and fine-tune the PLM with prompts to complete the FMSA task. Specifically, we make full use of the visual features with rich semantic information and modality sharing information obtained in the first stage through pre-training techniques. We use these features for lightweight and efficient parameter fine-tuning, successfully increasing the training efficiency while ensuring the predictive effect.

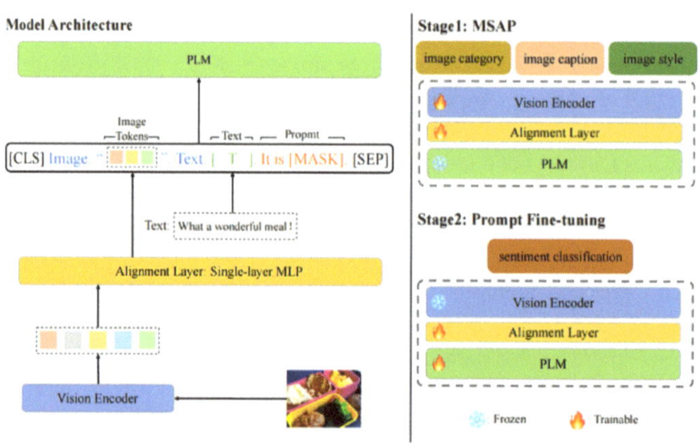

Figure 1. Our model architecture and training paradigm. The left side of the figure illustrates the unified model architecture for both training stages, while the right side of the figure depicts the two-stage training strategy and training tasks for each stage. "`Frozen`" indicates that a module's parameters remain unchanged during the training stage, whereas "`Trainable`" means the module's parameters will be updated in the training stage.

Our main contributions can be summarized as follows:

- We propose a novel Multi-task Supervised Alignment Pre-training (MSPA) framework to address the challenges in few-shot multimodal sentiment analysis (FMSA). This innovative approach significantly enhances modality alignment and improves the overall performance of few-shot multimodal sentiment analysis tasks.

- We design a multimodal supervised semantic alignment training paradigm that unifies diverse visual tasks, including image classification, image style recognition, and image captioning, into a cohesive Masked Language Model (MLM). This methodology enables the incorporation of rich semantic and modality-shared information into visual features, thereby facilitating more effective modality alignment.
- We integrate three distinct labeled image datasets to create a comprehensive corpus for our study. Extensive experimental evaluations conducted on this consolidated dataset demonstrate that our proposed method achieves state-of-the-art performance across three benchmark datasets for the FMSA task, substantiating the efficacy of our approach.

2. Related Work

In this section, we provide a brief overview of the related work of multimodal sentiment analysis, multi-task learning, and supervised pre-training.

Multimodal Sentiment Analysis. The applications of multimodal sentiment analysis are extensive, such as intelligent interactions, advertising marketing, and social media analysis [15]. Ref. [16] employed attention mechanisms and Recurrent Neural Networks (RNN) for sentiment analysis and mood detection. The attention mechanism can help the neural network focus on important parts of the sequence, thereby improving the performance of sentiment analysis and sentiment detection. Ref. [17] considered the necessity of uniformly processing text, audio, and image data; employed a deep Long Short-Term Memory (LSTM) model; and combined audio and visual modes for multimodal sentiment analysis. Ref. [18] established a model for information fusion in sentiment analysis tasks. The method takes into account the varying importance of information provided by different modalities for sentiment analysis. Ref. [19] proposed a deep multi-level attention network model for multimodal sentiment analysis. This network uses semantic attention to simulate the relationship between image and text modalities. Despite the sophistication of these methods, most of the above studies rely on large-scale annotated data, which consume a considerable cost. Therefore, we focus on efficiently using a small number of samples to achieve accuracy in sentiment analysis in an environment with limited resources. UP-MPF, as an advanced technology in this field, can well handle the text–image input and only requires pure unlabeled images rather than handcrafted text–image pairs to perform task-aligned multimodal pre-training for few-shot MSA. Based on this method, we use multi-task learning and supervised pre-training to solve the two main problems that UP-MPF currently faces.

Multi-task Learning. Multi-task learning is an advanced paradigm in machine learning that involves the simultaneous acquisition of knowledge from multiple related tasks. Its fundamental principle is to leverage shared information across tasks, thereby enhancing the model's generalization capabilities and overall performance [20]. Multi-task learning can effectively improve the performance of multimodal sentiment analysis [21]. In multimodal sentiment analysis, multi-task learning optimizes the training process through multiple tasks together, thereby achieving information complementarity between tasks. The different tasks of the model share the same base model, so different tasks can learn shared representations, extracting shared features of various modalities in multimodal sentiment analysis. When the model handles multiple tasks, the likelihood of model overfitting is reduced due to the increase in training data and constraints among multiple tasks. Ref. [22] jointly trains multimodal and unimodal tasks, learning consistency and differences, respectively. In addition, during the training stage, ref. [22] designed a weight adjustment strategy to balance the learning progress between different subtasks. Ref. [23] studied the concept of multi-task learning for multimodal sentiment analysis and explored a contextual multimodal attention framework. Ref. [23] defined three different multi-task settings, and each setting has two tasks. Therefore, we utilized multi-task learning, enabling the model to "synthesize" the knowledge learned from multiple tasks and preventing the model from overly focusing on any single task, which solves the second issue of UP-MPF. Specifically, in the classifying features of the MSAP stage, we chose to perform image classification tasks and image style recognition tasks; in addition, we added an image title matching task. These tasks can provide clearer semantic signals, enriching the embedded semantic information.

Supervised Pre-training. Supervised pre-training plays roles in promoting feature learning and transfer learning, enhancing system performance, and improving model robustness in multimodal sentiment analysis [24]. Supervised pre-training can learn high-order features and complex abstract patterns, which are vital for sentiment recognition. The application of supervised pre-training in multimodal sentiment analysis can indeed reduce the risk of model overfitting to a certain extent. In addition, the pre-trained model can be fine-tuned more effectively, thus improving the accuracy of sentiment analysis tasks. Ref. [25] performed supervised learning on the labels of multimodal tweets and designed three pre-training tasks that include text, visual, and multimodal elements. In order to conduct sentiment analysis, ref. [26] utilized as much multimodal information as possible. During the pre-training process, ref. [26] randomly masked the input tokens and then made predictions. Ref. [27] utilized the batch contrastive approach in the multimodal sentiment analysis system, transitioning from a fully supervised manner to a self-supervised one. Ref. [28] utilized domain-specific self-supervised graph convolution for multimodal sentiment analysis. By pre-training on a large amount of labeled data, the model can learn the shared features and patterns of multimodal data, providing strong support for subsequent sentiment analysis tasks. In addition, the pre-trained model can be transferred to specific sentiment analysis tasks, fully utilizing the knowledge learned by the model during the pre-training stage, which reduces the cost of starting training from scratch on a new task. Therefore, we address the lack of semantic signals in UP-MPF through supervised pre-training. Incorporating multi-task learning on the basis of UP-MPF can help the model better understand and utilize the correlation between different modal data, achieving better pattern alignment and improving model performance. Supervised pre-training can enhance representations, provide richer label information, and improve the model's generalization ability to enhance the learning effect of the UP-MPF algorithm on label information.

3. Methods

Task Definition. In this paper, we denote the multimodal sentiment analysis dataset as $D = (x, y)$. For the MSA task that mainly focuses on text and images, $x = (T, I)$, T denotes the sentence and I denotes the corresponding image that T is describing. $y \in Y$ denotes the predicted sentiment label, where Y is a predefined label set of sentiments (e.g., {positive, neutral, negative}). We assume access to a pre-trained language model M and use it to recognize the sentiment Y based on the image–text pair x. In the FMSA task, a limited number of training samples can be accessed. Additionally, the sizes of the development set and training set are equal, i.e., $|D_{\text{train}}| = |D_{\text{dev}}| \ll |D|$. This ensures consistency in the volumes of data used for creating the model and fine-tuning it, thereby giving us a fair framework to assess the effectiveness and efficiency of the model under similarly sized datasets.

3.1. Model Architecture

In this section, we will introduce our model in detail. Formally, given a template (e.g., Image: H_V. Text: T. It is [MASK].) that contains a text–image set and a designed text prompt with [MASK] token, we align the semantic information in the image to text through the alignment layer, then predict what may appear in [MASK], with an aim to help the model learn the multimodal sentiment information in both text and image forms.

As illustrated in Figure 1, the overall model architecture comprises a vision encoder f_V, a pre-trained foundation language model f_L, and a single-layer MLP as the alignment layer f_A. We employ the vision encoder and alignment layer to extract semantic information from the image and transform it into corresponding image tokens to align with the text T. For the image modality, we first employ NF-ResNet [29] as the vision encoder f_V to extract the vision representation X_V corresponding to the input image I. The visual representation X_V extracted at this stage forms a set of critical features that capture the essence of the visual elements presented in the image. Subsequently, we utilize the alignment layer f_A, a simple single-layer MLP, as a bridge that connects two different spaces: the vision embedding space and the language embedding space. The alignment layer f_A plays an instrumental role in connecting visual representation with linguistic representation. To achieve this, we employ a lightweight linear network as the alignment layer to obtain the vision–language representation of X_V, denoted as H_V, which can also be viewed as a sequence of image tokens. Formally, to obtain the image tokens H_V of the input image I,

$$X_V = f_V(I) \tag{1}$$

$$H_V = f_A(X_V) \tag{2}$$

where $H_V \in \mathbb{R}^{K \times d_L}$; d_L is the dimension of the language embedding space in the pre-trained language model; and K serves as a hyper-parameter, representing the length of the image token sequence and quantifying the extent of visual information encompassed within these tokens.

Furthermore, as a key component integrating visual and language understanding, the alignment layer f_A not only fulfills the crucial purpose of aligning modalities but also plays a pivotal role in effectively filtering out fine-grained visual information. It essentially performs a downsampling process, dimensionally reducing the larger image features produced within the vision encoder f_V and leading to a substantial reduction in computational complexity. Once we have obtained H_V, we combine the text T, the designed text prompt with [*MASK*] token (e.g., It is [*MASK*]), and the image tokens H_V to create a multimodal prompt template T. An example of T is provided below:

$$Image : H_V.Text : T. \; It \; is[MASK] \tag{3}$$

We then feed the embedding of T into the pre-trained language model M in order to predict the token with the highest probability at the [*MASK*] position. Additionally, we define a verbalizer, denoted as v, to convert sentiments into indices that point towards specific vocabulary. The process can be described as $Y \longrightarrow V$, where V is the vocabulary of M. For each label $y \in Y$, $v(y)$ maps the label to its corresponding token in V. Formally, obtaining the predicted label $\hat{y}$ given the image–text pair x can be illustrated as

$$\hat{y} = \arg\max \; p([\text{MASK}] = v(\hat{y}) \mid T(x)) \tag{4}$$

3.2. Multi-Task Supervised Alignment Pre-Training

To enhance the performance of multimodal prompt learning in the case of few shots, we utilize our MSAP method to promote the alignment of visual and language modes.

To enrich the semantic information embedded within vision features in the MSAP stage, we chose to perform the image classification task and the image style recognition task. As displayed in Table 1, the image classification task aims to classify the given images using a set of categories drawn from ImageNet. Concurrently, the image style recognition task is engaged in identifying the specific style that a given image exhibits. Additionally, we implemented an image caption matching task to improve visual–textual alignment. This task challenges the model to associate images with their correct captions, fostering a deeper understanding of cross-modal relationships. By including this task, we aim to enhance the model's ability to capture subtle semantic connections between modalities, ultimately improving its multimodal sentiment analysis performance. These tasks provide explicit semantic signals by offering information on image categories, styles, and fine-grained image content. As shown in Figure 1, to acquire more generalized modality-shared information, we unify these tasks into an MLM task for multi-task learning by employing prompts.

Table 1. Multimodal templates for pre-training tasks. H_V denotes the image tokens from the image input I. T denotes the caption text corresponding to the image I.

Task	Multimodal Prompt	Label Space
Image Classification	Image: H_V The class of the picture: [MASK]	{283 classes from ImageNet}
Image Caption Matching	If the picture: H_V matches the description: T? [MASK]	{yes, no}
Image Style Recognition	Image: H_V. The style of the picture: [MASK]	{real, sketch, painting}

To optimize the utilization of shared modal information, we devised a novel approach that employs a single MLM task to achieve the effects of multi-task learning. This methodology provides a scalable and unified multi-task learning paradigm, leveraging diverse prompt templates to encapsulate rich knowledge derived from multiple visual tasks within the learned image tokens. By consolidating varied visual objectives into a cohesive framework through carefully crafted prompts, we enable the model to distill and integrate complex visual information across various domains. This approach not only enhances the model's overall representational capacity and cross-modal understanding but also offers flexibility in task formulation through the strategic use of prompts, thereby facilitating more efficient and comprehensive learning. Table 1 presents the multimodal prompt templates set up for each task. While gravitating through the MSAP stage, the PLM is kept frozen and unaltered, ensuring its stability, thus exploiting its learned knowledge without any changes. By leveraging both the trainable vision encoder f_V and alignment layer f_A, this approach significantly enhances the ability to align modalities effectively. In simpler terms, the MSAP method facilitates the integration and alignment of semantic and shared modality information within visual features, thereby boosting the overall performance and effectiveness of the model across various tasks.

3.3. Downstream Prompt Fine-Tune

To better adapt PLM for the downstream task, it is vital to fine-tune the PLM. In the fine-tuning stage, we adopt an identical model architecture to that employed in the pre-training stage. We freeze the vision encoder f_V to obtain coherent vision–language representations. Simultaneously, we update the weights of the alignment layer f_A and PLM f_L to maximize its performance on the downstream task. The prompt we employed is as follows:

$$Image : H_V.Text : T. \ Sentiment \ of \ the \ text : [MASK]. \tag{5}$$

where H_V denotes the image tokens of the image input I and T denotes the corresponding text.

4. Experiments

4.1. Datasets and Data Split

Pre-training Dataset. We adopt different datasets for three tasks in the MSAP stage. For the image classification task, We choose 371,362 images from 283 categories in ImageNet [30]. For the image style recognition task, we filter 222,253 images from three styles in the Domain-Net dataset [31]. Additionally, we select 300,733 image–text pairs from the SBU Captions dataset [32] to design the image caption matching task. To ensure class balance for each task, we split them by a 9:1 ratio to form our training set and development set in the alignment pre-training stage.

Benchmark Datasets. We evaluate the effect of our MSAP method on three MSA datasets, including MVSA-S, MVSA-M [2], and TumEmo [3]. As for MVSA-S and MVSA-M, the composition of the two datasets is entirely consistent. Each row of the dataset contains an image and its corresponding text description, with its sentiment orientation labeled {Positive, Neutral, Negative}. As for the Tumemo dataset, although it maintains structural uniformity with the previous two, it provides sentiment labels, which contain {Angry, Bored, Calm, Fear, Happy, Love, Sad}. In light of the disparate volume of data associated with different label groups within the dataset, the application of sampling strategies becomes crucial in minimizing the impact of data imbalance. Therefore, we keep the test set unchanged and randomly sample 1% from the training and development sets to form our few-shot dataset (TumEmo is too large, so we only choose 497 samples). The statistics of the benchmark datasets are shown in Table 2.

Table 2. Statistics on the benchmark datasets.

Dataset	Train_{few}/Train	Dev_{few}/Dev	Test
MVSA-S	36/3224	36/403	408
MVSA-M	129/12,904	129/1613	1616
TumEmo	497/124,432	497/15,554	15,559

4.2. Implementation Details

We employ NF-Resnet50 as the vision encoder and BERT-base [33] as the PLM during both training stages. Additionally, we employ the early stopping strategy. During the MSAP stage, we optimize the model with a learning rate of 1×10^{-5}. To incorporate more visual information, we set the number of image tokens to seven. We select the checkpoint with the highest validation performance for the downstream task.

In the downstream FMSA stage, we use a learning rate of 5×10^{-5}. We conduct a greedy search on the number of image tokens from one to seven, allowing the model to find the most suitable level of visual information to incorporate. We run five times randomly for each experiment and discard the highest and lowest results, reporting only the middle-three average performance.

4.3. Experimental Results

In our experiment, we compare multiple multimodal model methods and further evaluate their performance according to two metrics: standard accuracy (Acc) and weighted F1 measure (Weighted-F1). Among other multimodal model methods, MVAN [3] and MGNNS [34] are multimodal fusion methods and LXMERT [35] is a classic pre-trained multimodal model. PF [36] and LM-BFF [37] employ prompt learning for texts. PVLM [5] directly feeds vision features into the PLM for multimodal prompt learning. UP-MPF [9], the previous state-of-the-art model, utilizes alignment pre-training to align modalities. From the result from our experiment, as shown in Table 3, we can conclude the following:

Table 3. Performance comparisons with standard deviations for few-shot experiments on three benchmark datasets.

Method	MVSA-S		MVSA-M		TumEmo	
	Acc	Weighted-F1	Acc	Weighted-F1	Acc	Weighted-F1
MVAN	42.77 (±3.18)	44.14 (±2.77)	46.16 (±8.22)	49.27 (±6.59)	30.29 (±1.78)	30.03 (±2.24)
MGNNS	34.40 (±2.09)	36.90 (±3.30)	40.03 (±3.01)	43.37 (±3.42)	41.29 (±0.40)	40.30 (±0.36)
LXMERT	47.27 (±1.79)	46.79 (±2.16)	49.08 (±3.04)	51.32 (±2.02)	40.44 (±0.77)	40.09 (±0.99)
LM-BFF	57.35 (±1.63)	58.45 (±2.05)	54.60 (±1.17)	57.15 (±1.21)	45.68 (±0.19)	45.72 (±0.18)
PF	55.60 (±1.99)	57.41 (±1.58)	55.39 (±0.79)	57.26 (±0.92)	47.03 (±0.32)	47.08 (±0.37)
PVLM	56.50 (±1.82)	57.89 (±1.05)	55.55 (±1.09)	57.68 (±0.74)	47.83 (±0.48)	47.84 (±0.75)
UP-MPF	58.21 (±2.55)	59.74 (±1.82)	58.10 (±1.08)	58.50 (±0.57)	48.27 (±0.25)	48.34 (±0.31)
MSAP	**64.87 (±0.50)**	**63.97 (±0.59)**	**62.75 (±0.38)**	**59.36 (±1.50)**	**49.48 (±0.19)**	**49.47 (±0.29)**

(1) The performance of prompt-based learning methods like PF and LM-BFF is consistently better than conventional multimodal fusion methods across the three datasets. This further proves the effectiveness of prompt-based methods, which may be attributed to the flexibility of prompt-based learning, as it can easily adapt to new tasks by tailoring prompts rather than requiring the extensive structural modifications that multimodal fusion methods may require.

(2) Methods like UP-MPF utilize vision–language alignment pre-training to help the model better understand the semantic information included in the image–text pair. This, in turn, enhances performance in the FMSA task compared to other methods, except our MSAP method.

(3) Our MSAP outperforms existing methods on all three datasets, achieving a new state-of-the-art performance on the FMSA task. Our MSAP approach distinguishes itself by garnering unique semantic information from a variety of tasks, along with the accomplishment of multimodal alignment between language and vision. This results in an enhanced performance in FMSA tasks as compared to the existing methods.

As shown in Table 3, our MSAP method outperforms existing methods on all three datasets, achieving a new state-of-the-art performance on the FMSA task.

In addition, as shown in Table 4, we comprehensively compare the performance of the MSAP method with classic baselines including UP-MPF on the fine-grained MSA task. We find that MSAP significantly outperforms existing methods on all three datasets, further verifying its effectiveness in this task.

Table 4. Performance comparisons with standard deviations for fine-grained experiments on three benchmark datasets.

Method	Twitter-15		Twitter-17		MASAD	
	Acc	Weighted-F1	Acc	Weighted-F1	Acc	Weighted-F1
TomBERT	44.89 (±3.97)	44.73 (±5.70)	47.37 (±0.93)	46.59 (±1.26)	73.21 (±1.06)	73.31 (±0.61)
EF-CaTrBERT	42.79 (±4.96)	41.43 (±8.64)	44.98 (±3.44)	44.34 (±2.67)	67.79 (±2.71)	68.29 (±2.69)
LXMERT	39.39 (±9.94)	37.78 (±7.01)	42.30 (±5.50)	36.30 (±2.20)	70.90 (±1.23)	71.42 (±1.17)
LM-BFF	49.68 (±3.62)	47.82 (±1.28)	50.76 (±0.84)	50.28 (±0.75)	74.67 (±0.51)	75.05 (±0.43)
PF	50.53 (±2.79)	50.14 (±4.22)	50.42 (±1.36)	49.06 (±4.28)	72.41 (±1.03)	72.66 (±0.82)
UP-MPF	56.03 (±3.15)	56.41 (±3.41)	52.61 (±2.04)	51.96 (±3.03)	75.82 (±0.59)	76.19 (±0.58)
MSAP	**64.67 (±0.55)**	**63.85 (±0.36)**	**62.86 (±0.59)**	**59.52 (±1.71)**	**79.62 (±0.36)**	**78.47 (±0.29)**

4.4. Ablation Experiment

4.4.1. Ablation of Fine-Tuning Strategies

Further investigation into fine-tuning strategies is necessary to maximize the utilization of these pre-trained models from our MSAP stage. As shown in Table 5, both fine-tuning the vision encoder and freezing the alignment layer can result in a decline in performance. Additionally, fine-tuning the vision encoder might further deteriorate the model's performance more than freezing the alignment layer does. When there are very little data available for downstream tasks, updating the visual encoder can potentially lead to severe overfitting, thereby distorting the representations learned from pre-training. However, updating the alignment layer is feasible since it has fewer parameters compared to the vision encoder. The above experiments demonstrate that freezing the visual encoder and fine-tuning the alignment layer is the optimal fine-tuning strategy, allowing for the acquisition of coherent vision–language features while transferring knowledge to downstream tasks.

Table 5. Performance comparisons of different training strategies for downstream tasks. f_V is the vision encoder, f_A is the alignment layer, "+" denotes the trainable module, and "−" denotes the frozen module.

Datasets	Strategy	Acc	Weighted-F1
MVSA-S	$f_V - f_A+$	**64.87 (±0.50)**	**63.97 (±0.59)**
	$f_V - f_A-$	52.21 (±6.92)	54.55 (±6.06)
	$f_V + f_A+$	52.37 (±0.31)	54.79 (±0.09)
	$f_V + f_A-$	49.43 (±0.64)	51.98 (±1.33)
MVSA-M	$f_V - f_A+$	**62.75 (±0.38)**	**59.36 (±1.50)**
	$f_V - f_A-$	56.77 (±4.05)	57.03 (±1.05)
	$f_V + f_A+$	60.64 (±2.36)	59.98 (±2.04)
	$f_V + f_A-$	60.54 (±5.31)	58.84 (±0.61)
TumEmo	$f_V - f_A+$	**49.48 (±0.19)**	**49.47 (±0.29)**
	$f_V - f_A-$	46.06 (±0.51)	45.75 (±0.59)
	$f_V + f_A+$	44.50 (±0.88)	44.60 (±1.05)
	$f_V + f_A-$	43.92 (±0.80)	43.17 (±0.55)

4.4.2. Ablation of Prompt Templates

To assess the impact of prompt template variations on model performance, as shown in Table 6, we conducted ablation experiments across coarse-grained, fine-grained, and MSA tasks. These experiments focused on modifying the prompt templates of downstream tasks, generating multiple variants to evaluate their effects.

Table 6. Performance comparisons of different prompt templates on downstream tasks.

Datasets	Prompt	Acc	Weighted-F1
MVSA-S	P1: Image: HV. Text: T. Sentiment of the text: [MASK].	64.87 (±0.52)	63.87 (±0.55)
	P2: Given the image HV and text T, the sentiment is [MASK].	64.83 (±0.48)	63.79 (±0.51)
	P3: For the image HV and the description T, the predicted sentiment is [MASK].	64.81 (±0.51)	63.80 (±0.48)
	P4: Analyze the image HV and the text T. The emotional tone is [MASK].	64.79 (±0.48)	63.83 (±0.52)
MVSA-M	P1: Image: HV. Text: T. Sentiment of the text: [MASK].	62.75 (±0.28)	59.56 (±1.20)
	P2: Given the image HV and text T, the sentiment is [MASK].	62.69 (±0.25)	59.36 (±1.10)
	P3: For the image HV and the description T, the predicted sentiment is [MASK].	62.71 (±0.23)	59.32 (±1.12)
	P4: Analyze the image HV and the text T. The emotional tone is [MASK].	62.69 (±0.21)	59.33 (±1.15)
TumEmo	P1: Image: HV. Text: T. Sentiment of the text: [MASK].	43.48 (±0.22)	49.43 (±0.31)
	P2: Given the image HV and text T, the sentiment is [MASK].	43.45 (±0.25)	49.39 (±0.29)
	P3: For the image HV and the description T, the predicted sentiment is [MASK].	43.44 (±0.24)	49.40 (±0.27)
	P4: Analyze the image HV and the text T. The emotional tone is [MASK].	43.41 (±0.25)	49.41 (±0.26)
Twitter-15	P1: Text: "S". Aspect: "A". Sentiment of aspect: [MASK].	63.48 (±0.22)	62.66 (±0.47)
	P2: Given the text "S" and the aspect "A", the associated sentiment is [MASK].	63.45 (±0.23)	62.67 (±0.45)
	P3: In the text "S", the sentiment related to the aspect "A" is [MASK].	63.42 (±0.21)	62.60 (±0.41)
	P4: Analyze the text "S". For the aspect "A", the detected sentiment is [MASK].	63.43 (±0.24)	62.63 (±0.44)

Table 6. *Cont.*

Datasets	Prompt	Acc	Weighted-F1
Twitter-17	P1: Text: "S". Aspect: "A". Sentiment of aspect: [MASK].	63.67 (±0.28)	60.56 (±1.29)
	P2: Given the text "S" and the aspect "A", the associated sentiment is [MASK].	63.65 (±0.27)	60.53 (±1.18)
	P3: In the text "S", the sentiment related to the aspect "A" is [MASK].	63.62 (±0.25)	60.48 (±1.11)
	P4: Analyze the text "S". For the aspect "A", the detected sentiment is [MASK].	6368 (±0.21)	60.55 (±1.21)
MASAD	P1: Text: "S". Aspect: "A". Sentiment of aspect: [MASK].	44.78 (±0.40)	51.03 (±0.35)
	P2: Given the text "S" and the aspect "A", the associated sentiment is [MASK].	44.75 (±0.42)	51.06 (±0.32)
	P3: In the text "S", the sentiment related to the aspect "A" is [MASK].	44.72 (±0.37)	51.11 (±0.21)
	P4: Analyze the text "S". For the aspect "A", the detected sentiment is [MASK].	44.77 (±0.39)	51.08 (±0.25)

Our findings indicate that while prompt design in few-shot learning influences performance, its overall impact remains limited. Despite structural variations, our approach exhibits strong robustness, enabling the model to effectively adapt to diverse downstream task requirements.

5. Conclusions

We propose a novel Multi-task Supervised Alignment Pre-training (MSAP) method for the few-shot multimodal sentiment analysis (FMSA) task. By incorporating richer semantic information and modality-shared information into vision features, MSAP further facilitates modality alignment, achieving a new state-of-the-art performance on three datasets for the FMSA task. In future work, we will explore more data-centric methods for alignment pre-training.

Author Contributions: Conceptualization, J.Y.; Methodology, J.Y.; Software, J.Y.; Validation, C.D.; Resources, C.D.; Data curation, C.D.; Writing—original draft, J.Y.; Writing—review & editing, J.C. All authors have read and agreed to the published version of the manuscript.

Funding: This research received no external funding.

Institutional Review Board Statement: Not applicable.

Informed Consent Statement: Not applicable.

Data Availability Statement: Data is contained within the article. The data presented in this study are available in Section 4.1 of the article.

Conflicts of Interest: The funders had no role in the design of the study; in the collection, analyses, or interpretation of data; in the writing of the manuscript; or in the decision to publish the results.

References

1. Kaur, R.; Kautish, S. Multimodal sentiment analysis: A survey and comparison. In *Research Anthology on Implementing Sentiment Analysis Across Multiple Disciplines*; IGI Global: Hershey, PA, USA, 2022; pp. 1846–1870.
2. Niu, T.; Zhu, S.; Pang, L.; El Saddik, A. Sentiment analysis on multi-view social data. In Proceedings of the MultiMedia Modeling: 22nd International Conference, MMM 2016, Miami, FL, USA, 4–6 January 2016; Springer: Berlin/Heidelberg, Germany, 2016; Proceedings, Part II 22; pp. 15–27.
3. Yang, X.; Feng, S.; Wang, D.; Zhang, Y. Image-text multimodal emotion classification via multi-view attentional network. *IEEE Trans. Multimed.* **2020**, *23*, 4014–4026. [CrossRef]
4. Liu, P.; Yuan, W.; Fu, J.; Jiang, Z.; Hayashi, H.; Neubig, G. Pre-train, prompt, and predict: A systematic survey of prompting methods in natural language processing. *Acm Comput. Surv.* **2023**, *55*, 1–35. [CrossRef]
5. Yu, Y.; Zhang, D. Few-shot multi-modal sentiment analysis with prompt-based vision-aware language modeling. In Proceedings of the 2022 IEEE International Conference on Multimedia and Expo (ICME), Taipei, Taiwan, 18–22 July 2022; pp. 1–6.
6. Zhu, D.; Chen, J.; Shen, X.; Li, X.; Elhoseiny, M. Minigpt-4: Enhancing vision-language understanding with advanced large language models. *arXiv* **2023**, arXiv:2304.10592.
7. Liu, H.; Li, C.; Wu, Q.; Lee, Y.J. Visual instruction tuning. *Adv. Neural Inf. Process. Syst.* **2024**, *36*.
8. Ye, Q.; Xu, H.; Xu, G.; Ye, J.; Yan, M.; Zhou, Y.; Wang, J.; Hu, A.; Shi, P.; Shi, Y.; et al. mplug-owl: Modularization empowers large language models with multimodality. *arXiv* **2023**, arXiv:2304.14178.

9. Yu, Y.; Zhang, D.; Li, S. Unified multi-modal pre-training for few-shot sentiment analysis with prompt-based learning. In Proceedings of the 30th ACM International Conference on Multimedia, Lisbon, Portugal, 10–14 October 2022; pp. 189–198.
10. Gidaris, S.; Singh, P.; Komodakis, N. Unsupervised representation learning by predicting image rotations. *arXiv* **2018**, arXiv:1803.07728.
11. Liu, Y. Multilingual denoising pre-training for neural machine translation. *arXiv* **2020**, arXiv:2001.08210. [CrossRef]
12. Feng, Y.; Jiang, J.; Tang, M.; Jin, R.; Gao, Y. Rethinking supervised pre-training for better downstream transferring. *arXiv* **2021**, arXiv:2110.06014.
13. Collobert, R.; Weston, J. A unified architecture for natural language processing: Deep neural networks with multitask learning. In Proceedings of the 25th International Conference on Machine Learning, Helsinki, Finland, 5–9 July 2008; pp. 160–167.
14. He, H.; Choi, J.D. The stem cell hypothesis: Dilemma behind multi-task learning with transformer encoders. *arXiv* **2021**, arXiv:2109.06939.
15. Gandhi, A.; Adhvaryu, K.; Poria, S.; Cambria, E. Hussain Multimodal sentiment analysis: A systematic review of history, datasets, multimodal fusion methods, applications, challenges and future directions. *Inf. Fusion* **2023**, *91*, 424–444. [CrossRef]
16. Wang, R.; Geng, Z. Wa-icp algorithm for tackling ambiguous correspondence. In Proceedings of the IEEE 2015 3rd IAPR Asian Conference on Pattern Recognition (ACPR), Kuala Lumpur, Malaysia, 3–6 November 2015; pp. 76–80.
17. Sindhu, P. The future of transport networks. In Proceedings of the Optical Fiber Communication Conference, Los Angeles, CA, USA, 22–26 March 2015; Optica Publishing Group: Washington, DC, USA, 2015.
18. Wang, B.; Dong, G.; Zhao, Y.; Li, R.; Cao, Q.; Chao, Y. Non-uniform attention network for multi-modal sentiment analysis. In *International Conference on Multimedia Modeling*; Springer: Berlin/Heidelberg, Germany, 2022; pp. 612–623.
19. Yadav, A.; Vishwakarma, D.K. A deep multi-level attentive network for multimodal sentiment analysis. *ACM Trans. Multimed. Comput. Commun. Appl.* **2023**, *19*, 1–19. [CrossRef]
20. Ruder, S. An overview of multi-task learning in deep neural networks. *arXiv* **2017**, arXiv:1706.05098.
21. Yang, B.; Wu, L.; Zhu, J.; Shao, B.; Lin, X.; Liu, T.-Y. Multimodal sentiment analysis with two-phase multi-task learning. *IEEE/ACM Trans. Audio Speech Lang. Process.* **2022**, *30*, 2015–2024. [CrossRef]
22. Yu, W.; Xu, H.; Yuan, Z.; Wu, J. Learning modality-specific representations with self-supervised multi-task learning for multimodal sentiment analysis. *AAAI Conf. Artif. Intell.* **2021**, *35*, 10790–10797. [CrossRef]
23. Akhtar, M.S.; Chauhan, D.S.; Ekbal, A. A deep multi-task contextual attention framework for multi-modal affect analysis. *ACM Trans. Knowl. Discov. Data* **2020**, *14*, 1–27. [CrossRef]
24. Zadeh, A.B.; Liang, P.P.; Poria, S.; Cambria, E.; Morency, L.-P. Multimodal language analysis in the wild: Cmu-mosei dataset and interpretable dynamic fusion graph. In Proceedings of the 56th Annual Meeting of the Association for Computational Linguistics, Melbourne, Australia, 15–20 July 2018; Volume 1: Long Papers, pp. 2236–2246.
25. Ling, Y.; Yu, J.; Xia, R. Vision-language pre-training for multimodal aspect-based sentiment analysis. *arXiv* **2022**, arXiv:2204.07955.
26. Kim, K.; Park, S. Aobert: All-modalities-in-one bert for multimodal sentiment analysis. *Inf. Fusion* **2023**, *92*, 37–45. [CrossRef]
27. Mai, S.; Zeng, Y.; Zheng, S.; Hu, H. Hybrid contrastive learning of tri-modal representation for multimodal sentiment analysis. *IEEE Trans. Affect. Comput.* **2022**, *14*, 2276–2289. [CrossRef]
28. Zeng, Y.; Li, Z.; Tang, Z.; Chen, Z.; Ma, H. Heterogeneous graph convolution based on in-domain self-supervision for multimodal sentiment analysis. *Expert Syst. Appl.* **2023**, *213*, 119240. [CrossRef]
29. Brock, A.; De, S.; Smith, S.L. Characterizing signal propagation to close the performance gap in unnormalized resnets. *arXiv* **2021**, arXiv:2101.08692.
30. Deng, J.; Dong, W.; Socher, R.; Li, L.-J.; Li, K.; Li, F.F. Imagenet: A large-scale hierarchical image database. In Proceedings of the 2009 IEEE Conference on Computer Vision and Pattern Recognition, Miami, FL, USA, 20–25 June 2009; pp. 248–255.
31. Peng, X.; Bai, Q.; Xia, X.; Huang, Z.; Saenko, K.; Wang, B. Moment matching for multi-source domain adaptation. In Proceedings of the IEEE/CVF International Conference on Computer Vision, Seoul, Republic of Korea, 27 October–2 November 2019; pp. 1406–1415.
32. Ordonez, V.; Kulkarni, G.; Berg, T. Im2text: Describing images using 1 million captioned photographs. *Adv. Neural Inf. Process. Syst.* **2011**, *24*.
33. Devlin, J. Bert: Pre-training of deep bidirectional transformers for language understanding. *arXiv* **2018**, arXiv:1810.04805.
34. Yang, X.; Feng, S.; Zhang, Y.; Wang, D. Multimodal sentiment detection based on multi-channel graph neural networks. In Proceedings of the 59th Annual Meeting of the Association for Computational Linguistics and the 11th International Joint Conference on Natural Language Processing, Online, 1–6 August 2021; Volume 1: Long Papers, pp. 328–339.
35. Tan, H.; Bansal, M. Lxmert: Learning cross-modality encoder representations from transformers. *arXiv* **2019**, arXiv:1908.07490.

36. Schick, T.; Schütze, H. Exploiting cloze questions for few shot text classification and natural language inference. *arXiv* **2020**, arXiv:2001.07676.
37. Gao, T.; Fisch, A.; Chen, D. Making pre-trained language models better few-shot learners. *arXiv* **2020**, arXiv:2012.15723.

applied sciences

MDPI

Article

Evaluating Retrieval-Augmented Generation Models for Financial Report Question and Answering

Ivan Iaroshev [1], Ramalingam Pillai [1], Leandro Vaglietti [1] and Thomas Hanne [2,*]

1 School of Business, University of Applied Sciences and Arts Northwestern Switzerland, 4600 Olten, Switzerland
2 Institute for Information Systems, University of Applied Sciences and Arts Northwestern Switzerland, 4600 Olten, Switzerland
* Correspondence: thomas.hanne@fhnw.ch; Tel.: +41-62-957-22-92

Abstract: This study explores the application of retrieval-augmented generation (RAG) to improve the accuracy and reliability of large language models (LLMs) in the context of financial report analysis. The focus is on enabling private investors to make informed decisions by enhancing the question-and-answering capabilities regarding the half-yearly or quarterly financial reports of banks. The study adopts a Design Science Research (DSR) methodology to develop and evaluate an RAG system tailored for this use case. The study conducts a series of experiments to explore models in which different RAG components are used. The aim is to enhance context relevance, answer faithfulness, and answer relevance. The results indicate that model one (OpenAI ADA and OpenAI GPT-4) achieved the highest performance, showing robust accuracy and relevance in response. Model three (MiniLM Embedder and OpenAI GPT-4) scored significantly lower, indicating the importance of high-quality components. The evaluation also revealed that well-structured reports result in better RAG performance than less coherent reports. Qualitative questions received higher scores than the quantitative ones, demonstrating the RAG's proficiency in handling descriptive data. In conclusion, a tailored RAG can aid investors in providing accurate and contextually relevant information from financial reports, thereby enhancing decision making.

Keywords: retrieval-augmented generation; large language models (LLMs); financial reports; question and answering with LLM

1. Introduction

Since the inception of Artificial Intelligence (AI), researchers have pursued the ambitious goal of developing machines capable of reading, writing, and conversing like humans. This occurred in the domain of natural language processing (NLP). According to Zhao et al. [1], NLP has a rich history: from the development of statistical language models (SLMs) to the rise of neural language models (NLMs), linguistic capabilities have significantly increased, enabling AI systems to understand and generate complex language patterns [1].

In recent years, NLP has witnessed remarkable progress with the introduction of large language models (LLMs). These models, which were extensively trained on vast amounts of textual data, demonstrated unprecedented proficiency in generating human-like text and performing language-based tasks accurately [2]. However, despite their advancements, LLMs still face several limitations. Burtsev et al. [3] outline three main limitations of LLMs. First, they struggle with complex reasoning, hindering their ability to draw accurate conclusions. Second, their knowledge or expertise is limited to training data, which may lead to failure in providing relevant information. Third, they may produce inaccurate outputs due to a lack of understanding of prompts, which makes them marginally helpful.

The rise of LLMs has also sparked interest in their application to specific tasks, prompting the emergence of an approach called retrieval-augmented generation (RAG). This approach was first introduced by Guu et al. [4] and Lewis et al. [5]. RAG was devised to extend the capabilities of LLMs beyond conventional training data. By integrating a specialized body of knowledge, RAG enables LLMs to provide more accurate responses to user queries. In essence, RAG comprises two distinct phases: retrieval and generation. During the retrieval phase, defined sources (e.g., indexed documents) are analyzed to extract relevant information that is aligned with the user's prompt or question. The retrieved information is then seamlessly integrated with the user's prompt and forwarded to the language model. In the generative phase, the LLM leverages the augmented prompt and internal understanding to craft a tailored response that addresses the user's query effectively [6,7].

RAG is gaining momentum for its potential to improve LLM-generated responses by grounding models on the said external sources, thereby reducing issues such as inconsistency and hallucination [8,9].

1.1. Problem Statement

The integration of RAG into LLMs has rapidly gained traction, and it has emerged as a key strategy to increase the practical applicability of LLMs [8]. However, despite its potential to reduce factual inaccuracies or "hallucinations", there remains a notable gap in the literature regarding RAG's application in question and answering for a specific use case: private investors seeking to analyze the financial reports of banks effectively (see Section 2.4). The manual analysis of such reports is both time-consuming and labor-intensive. However, by leveraging RAG, investors can access domain-specific information from half-yearly or quarterly reports previously published by several banks. Through RAG-facilitated question-and-answer sessions, investors can make informed decisions based on accurate insights.

This study focuses on half-yearly and quarterly reports due to their regular publication frequency, ensuring an ample number of samples for analysis. In addition, these reports contain a better-manageable number of pages than annual reports, which may overwhelm an RAG system with their extensive content. Further details about the data collection process are provided in Section 3.2.

Currently, there is a lack of a straightforward RAG approach tailored to this use case, which hinders the empirical demonstrations of how RAG can be effectively adapted (Section 2.3). These experiments with different model configurations should focus on both the retrieval and generation components (Section 2.2) to achieve desirable outcomes across evaluation aspects (context relevance, answer faithfulness, and answer relevance, Section 2.3).

Therefore, the motivation to research RAG encompasses several crucial aspects. First, it aims to enhance the quality of responses across evaluation dimensions such as context relevance, answer faithfulness, and answer relevance by exploring different model configurations (e.g., different LLMs enriched with the same data sources). Second, it attempts to address the persistent challenges encountered by state-of-the-art LLMs, such as factual inaccuracies and hallucinations. By examining the practical implementation and effectiveness of RAG, particularly in specific domains like financial analysis for private investors, this research aims to provide valuable insights that bridge the gap between theoretical advancements and real-world applicability.

1.2. Thesis Statement and Research Questions

Our study aims to explore the potential of RAG for improving the accuracy and reliability of LLMs for analyzing half-yearly or quarterly reports. Through a range of experiments featuring various RAG model configurations, such as different LLMs and retrieval methods, we aim to examine and gauge differences and improvements in context relevance, answer faithfulness, and answer relevance in financial report analysis. Ultimately, the goal is to empower users to make informed decisions by providing accurate answers. To achieve this, a thesis statement and main research question were formulated, supplemented by additional sub-questions.

Thesis Statement: The effectiveness of RAG in enhancing context relevance, answer faithfulness, and answer relevance for analyzing half-yearly or quarterly reports can be evaluated through empirical experiments with different model configurations.

Main Research Question: How can RAG be adapted to increase the context relevance, answer faithfulness, and answer relevance of LLM conclusions about several types of financial reports?

1. How can the retrieval component of RAG be set up to extract relevant information from half-yearly or quarterly reports?
2. What strategies can be employed to ensure that the generation component of RAG produces answers to the questions posed about half-yearly or quarterly reports?
3. How can the effectiveness of the RAG system for analyzing half-yearly or quarterly reports in the banking sector be reliably evaluated and validated?
4. How accurately do RAG-generated responses represent the information extracted from the half-yearly or quarterly reports of banking institutions?

2. Literature Review

In this review, RAG systems, including evaluation metrics, methods, and frameworks, are briefly discussed. In addition, a research gap is identified that requires further investigation.

Methodologically, the literature review was initiated with extensive searches on Google Scholar using broad keywords such as "Retrieval Augmented Generation" and "RAG" to capture general papers on the topic. As the investigation progressed, the search terms were refined to focus on specific aspects of RAG, including "Evaluation methods of RAG". This approach includes both forward and backward searches to ensure the comprehensive coverage of the relevant literature. While priority was given to peer-reviewed academic papers, a selection of non-peer-reviewed papers with substantial citations and reputable authors was also integrated, acknowledging the rapid advancements in LLM and RAG research.

2.1. History and Current State of RAG

The field of NLP has seen significant advancements with the emergence of LLMs like the GPT and LLama series, as well as models like Gemini [2,10,11]. Despite their successes, these models still need to cope with challenges such as outdated knowledge and a lack of domain-specific expertise [12,13]. One major issue is the generation of incorrect information, known as "hallucinations", especially when faced with queries beyond their training data [3,14]. These limitations highlight the need for additional safeguards before deploying LLMs in real-world settings [6].

A solution to address these challenges is RAG, which integrates external data retrieval into the generative process. RAG, pioneered by Lewis et al. [5] and Guu et al. [4], revolutionizes generative tasks within the LLM domain. In RAG, LLMs initially query external data sources to gather relevant information before generating text or answering questions. This retrieval step ensures that the subsequent generation phase is grounded in the retrieved evidence, which significantly improves the accuracy and relevance of the output [8,15].

By dynamically retrieving information from knowledge bases during inference, RAG effectively tackles issues such as factual inaccuracies or "hallucinations" [9]. The adoption of RAG in LLMs has rapidly gained momentum, and it has become a central approach to enhance the capabilities of chatbots and make LLMs more suitable for practical applications [8].

2.2. Evaluating Targets of RAG

Despite the proven effectiveness of RAG as outlined in Section 2.1, their successful implementation still requires careful tuning. Various factors, including the retrieval model, construction of the external knowledge base, and language model complexities, collectively influence the overall performance of the RAG system. Therefore, the evaluation of RAG systems is critical [16]. However, before proceeding with evaluation, it is important to define evaluation targets.

The workflow of a traditional RAG typically comprises three key steps according to Gao et al. [6]: indexing, retrieval, and generation. Indexing involves extracting raw data from various formats, converting the data to plain text, segmenting the data into smaller chunks, and encoding these chunks into vector representations. Vector representations are the numerical representations of data, such as words or documents, that are used to capture semantic meaning and enable computational operations. During retrieval, user queries are transformed similarly into vector representations using the same encoding model applied during indexing. The query vectors are then compared with the vector representations of the indexed chunks to compute similarity scores. The chunks with the highest similarity scores are retrieved and serve as the expanded context for generating responses. In the generation phase, the query and selected chunks are synthesized as a prompt for an LLM to formulate a response. This process allows the model to leverage the encoded representations of the input data to generate contextually relevant and coherent responses [6].

Let us note that there are slightly different approaches for RAG conceptualizations such as in IBM Research [7] where only the two phases retrieval and content generation are distinguished. However, both approaches agree that evaluation should ensure a comprehensive assessment of both the quality of the retrieved context and the coherence of the generated content. The retrieval quality assesses the effectiveness of the sourced context, while the generation quality evaluates the relevance and coherence of the responses [7].

2.3. Evaluation Aspects of RAG

In the evaluation of RAG models, there are three primary quality metrics to assess the two main targets retrieval and generation. These quality metrics, including context relevance (retrieval metric), answer faithfulness (generation metric), and answer relevance (generation metric), serve to gauge the effectiveness of the RAG system [17,18].

Context relevance scrutinizes the precision and specificity of the retrieved context, ensuring relevance while minimizing the processing costs associated with irrelevant content. Answer faithfulness ensures the fidelity of the generated responses to the retrieved context, thereby maintaining consistency and avoiding contradictions and "hallucinations". Answer relevance requires that the generated responses directly address the posed questions, effectively tackling the main inquiry [17,18].

RAG evaluation methods can be categorized into two types: reference-required and reference-free evaluation. In the reference-free approach, the quality of outputs is determined based on intrinsic text characteristics, bypassing the need for human-annotated ground truth labels in the evaluation dataset. Conversely, the reference-required approach compares outputs to a ground truth annotated by humans [16].

Reference-free evaluation frameworks leverage LLMs to assess quality scores, such as context relevance, answer faithfulness, and answer relevance, by analyzing the alignment of the generated text with the retrieved context. However, this method's reliability can be compromised if the retrieved external information is of low quality [16]. Frameworks such as RAGAS (Retrieval-Augmented Generation Assessment) by Es et al. [17,19], ARES (Automated RAG Evaluation System) by Saad-Falcon et al. [18], or RAG Triad by TruLens [20] are classified as reference-free frameworks according to [16]. Nonetheless, both RAGAS and ARES require a form of "ground truth" for evaluation besides using synthetic training data for assessing the quality of individual RAG components. Let us note that various other approaches related to reference-free evaluation methods are available [21–25], which, for instance, may require the manual evaluation of RAG's output.

In contrast to reference-free evaluation methods, reference-required benchmarks involve comparing outputs to a reference or ground truth annotated by humans. These benchmarks provide quantitative metrics to measure the performance of the RAG model and deepen the understanding of its capabilities across various evaluation aspects. For instance, the Retrieval-Augmented Generation Benchmark (RGB) [8] evaluates four essential abilities for RAG: noise robustness, negative rejection, information integration, and counterfactual robustness. However, this approach focuses on question-and-answering tasks to measure RAG performance. Other approaches, such as those discussed by Lyu et al. [16], support a wider range of RAG applications but are more challenging to implement.

2.4. RAG as an Approach to Financial Report Question and Answering

RAG finds application in scenarios centered around evaluating the question and answering of financial reports, offering access to a wealth of freely available domain-specific information, such as reports of banks. However, there is limited literature on this topic.

One approach targets financial analysts: Sarmah et al. [26] demonstrated the use of LLMs to efficiently extract information from earnings report transcripts while ensuring high accuracy. Combining the retrieval-augmented generation technique with metadata, they streamline the extraction process and mitigate hallucination. Their evaluation, based on various objective metrics for Q&A systems, demonstrated the superiority of their method. However, they focus on earnings call transcripts. Additionally, their approach incorporates manual labeling to evaluate the answer quality. Therefore, their evaluation methodology focuses on reference-required evaluation (as discussed in Section 2.3).

Two other papers explore very specific and narrow aspects of RAG in financial contexts. Yepes et al. [27] focused on effective chunking in RAG and proposed an enhanced approach to chunk financial reports by identifying structural elements in such documents. The proposed method optimizes the chunk size without tuning and improves context and accuracy.

Another important angle is financial sentiment analysis, which is crucial for valuation and investment decision making. Zhang et al. [28] introduced a retrieval-augmented LLM framework for this purpose, which features an instruction-tuned LLM module. This module uses pretrained LLMs with human-like text to guide the model's execution based on task descriptions and desired outputs, which are often labeled by humans. In addition, a retrieval-augmentation module gathers context from reliable external sources. However, this approach focuses on sentiment analysis rather than Q&A from financial reports.

2.5. Research Gap

In Sections 2.1–2.5, RAG applications were discussed that can effectively tackle the challenges of LLMs, such as inaccuracies and hallucinations. However, there is a significant lack of empirical knowledge and literature on optimizing RAG systems and their retrieval and generation components for financial report question and answering. Specifically, there is insufficient empirical evidence on the quality of RAG systems developed to handle complex financial data, such as quarterly and half-yearly bank reports.

Although automatic evaluation methods like RAGAS, as well as manual evaluation methods, exist to assess RAG systems without heavily relying on ground truths, there is limited literature on evaluating RAG systems specifically for bank reports. These reports pose unique challenges, such as accurately interpreting financial data, understanding regulatory compliance, and assessing business performance metrics. Therefore, evaluating RAG systems and their components in such scenarios to ensure minimal deviations in interpretation requires a thorough examination of the retrieval and generation components.

3. Research Design

This section presents the research methodology used to address the research questions outlined in Section 1.2 and delineates the data sources and collection methods employed.

The research tackles a real-world challenge by developing a novel and innovative artifact. Hence, Design Science Research (DSR), as proposed by Hevner and Chatterjee [29], is recommended. DSR involves the creation and evaluation of an artifact. In this context, it entails the establishment of an RAG system, along with the adaptation of its components, such as the embedding model or LLM. These adaptations will be tailored to varying model configurations, which will be evaluated. This application aims to tackle an important problem by addressing the lack of a straightforward RAG approach tailored for the use case of private investors analyzing half-yearly or quarterly bank reports.

3.1. Process Steps of Design Science Research

This research adheres to the 5-step approach of DSR as proposed by Hevner and Chatterjee [29]. The first step, Awareness, involves identifying the research problem and requirements. Subsequently, the second and third steps, Suggestion and Development, involve designing an artifact that addresses the identified problems. Following the design phase, the fourth step, Evaluation, assesses the effectiveness and functionality of the developed artifact. Finally, the fifth step, Conclusion, summarizes by synthesizing the findings and drawing conclusions.

Most steps of the DSR approach focus on a specific research question and its corresponding objective. This will be outlined in the following subsections.

3.1.1. Awareness of Problem and Identification of Requirements

In the initial step of the 5-step DSR approach proposed by Hevner and Chatterjee [29], the focus is on recognizing a problem that can be addressed through new artifacts. This step corresponds with the problem statement outlined in Section 1.1, the literature review in Section 2, and the research gap identified in Section 2.5. In short, the problem statement and research gap illustrate the limited exploration of RAG for assisting private investors in analyzing the financial reports of banks. Despite these potential benefits, tailored RAG approaches for this specific case are lacking. This gap hinders the empirical demonstration of RAG's effectiveness in providing accurate insights for investors. The motivation behind this research is to develop and test RAG implementations tailored to the analysis of the half-yearly and quarterly reports of banks with the aim of enhancing response quality, addressing challenges in dialog models, and contributing to advancing RAG applications in this domain.

3.1.2. Suggestion

The second step of DSR involves determining the type of artifact that could solve the current problem [29]. In this case, the proposed artifact is an RAG system and its corresponding implementation. This RAG system should be designed to facilitate the analysis of the half-yearly and quarterly reports of banks. It is essential to note that the RAG system, respectively, the retrieval and generation components, should be able to encompass various scenarios, thereby allowing for comprehensive evaluation and adaptation as needed.

3.1.3. Development

After defining the problem and identifying the artifact type, as per Hevner and Chatterjee [29], the Development phase should concentrate on designing and creating an artifact that offers a solution to the defined problem. The sub-questions one and two (see Table 1), outlined in Section 1.2., are integral to this phase.

Table 1. Sub-questions for the Development phase.

No.	Sub-Questions	Activities
1.	How can the retrieval component of RAG be set up to extract relevant information from half-yearly or quarterly reports?	Data collection through literature review Artifact development
2.	What strategies can be employed to ensure that the generation component of RAG produces answers to questions posed about half-yearly or quarterly reports?	Data collection through literature review Artifact development

In this phase, the focus is on developing and implementing an RAG system. Establishing an RAG system requires the definition and selection of certain components. Figure 1 shows the simplified architecture.

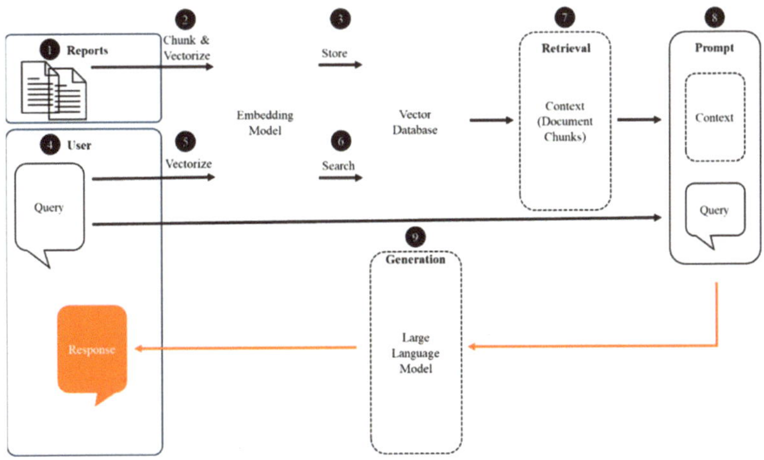

Figure 1. RAG system according to [6].

The flow is outlined based on the following 9 steps (identified by the circled number in Figure 1) based on [6]:

1. Begin by gathering essential reports for the RAG system's operation, specifically targeting half-yearly and quarterly reports from banks. Further insights into the data collection procedures are provided in Section 3.2.
2. Break down the collected data into manageable chunks to streamline information retrieval and improve efficiency by avoiding the processing of entire documents. This segmentation ensures that each data chunk remains focused on a specific topic, which increases the likelihood of retrieving relevant information for user queries. Subsequently, these segmented data are transformed into vector representations known as embeddings, which capture the semantic essence of the text.
3. The resulting embeddings are stored in a dedicated vector database to facilitate the efficient retrieval of pertinent information, transcending traditional word-to-word comparison methods.
4. The user query formulation is initiated.
5. Upon entry into the system, the user query is converted to an embedding or vector representation. To ensure consistency, the same model is used for both document and query embeddings.
6. Use the transformed query to search the vector database and perform comparisons with document embeddings.
7. The most relevant text chunks are retrieved, and a contextual framework is established to address the user query.
8. The retrieved text chunks are integrated with the original user query to provide a unified prompt for LLM.
9. The unified prompt, which comprises the retrieved text chunks and the original user query, is sent to the LLM. The LLM utilizes its advanced natural language processing capabilities and extensive knowledge to generate coherent responses tailored to address the user's queries effectively, leveraging the additional context provided by the text chunks to enhance the accuracy and depth of the responses.

Therefore, for the development of a newly built RAG system, the essential components include an embedding model, a vector database, an LLM, and a comprehensive orchestration tool to manage the entire system seamlessly. The following tools were identified as potential setups:

- Embedding Model: MiniLMEmbedder [30] and BedRockEmbeddings [31];
- Vector Database: FAISS [32] and Chroma [33];
- LLM: GPT-4o [34], Llama 3 [35], and Gemini 1.5 Pro [36];
- Orchestrator: LangChain [37] and LlamaIndex [38].

Furthermore, there exists an alternative approach involving the use of a pre-built RAG system that can be readily deployed. One such system is the Verba RAG by Weaviate. In this case, it is also possible to create different model configurations by using different embedding models (e.g., OpenAI ADA, MiniLMEmbedder) and LLMs (e.g., GPT-4o, Gemini 1.5 Pro) [39,40].

In any case, it is the goal to create three distinct technical model configurations, each using different components. For each model configuration, a selection of quarterly and half-yearly reports is gathered from various banks, as detailed in Section 3.2. Using these documents, querying and answering via the RAG system will be conducted. Ten questions per bank are formulated for this purpose and tested with each model configuration.

3.1.4. Feedback and Evaluation

In the fourth step of the DSR approach, the focus shifts to evaluating whether the artifact effectively addresses the identified problem and assessing its strengths and weaknesses [29]. In this context, the sub-questions 3 and 4 from Section 1.2. are relevant (see Table 2).

Table 2. Sub-questions for the Feedback and Evaluation phase.

No.	Sub-Questions	Activities
3.	How can the effectiveness of the RAG system for analyzing half-yearly or quarterly reports in the banking sector be reliably evaluated and validated?	The designing, conduction, and evaluation of tests
4.	How accurately do RAG-generated responses represent the information extracted from half-yearly or quarterly reports of banking institutions?	The designing, conduction, and evaluation of tests

The evaluation (sub-question 3 and 4) is crucial for determining the viability of an artifact in practical application scenarios, as it provides insights into the quality of the RAG system's performance. Three model configurations are generated for the subsequent evaluation. The evaluation of the answers provided by the RAG system encompasses context relevance, answer faithfulness, and answer relevance, as discussed in Section 2.3.

Given the considered use case, practical and accessible evaluation methods are needed that do not require extensive technical expertise. Due to the project's timeline and scope, we will set up an evaluation manually.

3.1.5. Conclusions

In the final step of the DSR process, the focus is on discussing the findings and evaluating their generalizability, as well as considering future aspects, such as open questions and plans for further development [29].

Thus, the aim is to illuminate the potential of the specific RAG system for broader implementation, identify avenues for continued development and enhancement, and address any open questions regarding its functionality, model configuration, output, or evaluation.

3.2. Data Collection and Analysis

This section discusses the necessary activities and the chosen data collection approach. As mentioned in Section 1.1, the half-yearly and quarterly reports of banks were used to evaluate the RAG system in combination with qualitative methods like document reviews by directly checking the considered quarterly and half-yearly reports. The reporting periods include the first quarters of 2022 and 2023 and the interim reports of the first half of 2022 and 2023. These periods were chosen because the data in these reports are sufficiently up-to-date. Another reason for the report's types of choice is that we would like to conduct and assess RAG performance under different conditions. Thus, the financial situation of each bank is represented by two half-yearly reports from 2022 and 2023 and two quarterly reports from the aforementioned years.

The chosen sample banks include Barclays, HSBC, Credit Suisse, Credit Agricole, and Banco Santander. The reasons for choosing these banks are diverse. First, we need to obtain direct access to their reports of the considered periods, which are published officially and are written in English. Second, the sample banks are from European countries with local economic features: the UK, EU (France, Austria, Spain, and Czech Republic), and Switzerland. It is well known that the British economic system is quite different from the EU's because of the domination of English law, which establishes other rules on property, trade, etc. At the same time, these three economic systems are not too far from each other unlike the US and China, which have varying standards of financial reporting. However, three additional banks (Bawag Bank, Erste Bank & Sparkasse, and KB Bank) were chosen. These extra documents are not relevant to the experiments; however, they test the RAG system's ability to filter out irrelevant data and focus on important information.

4. Solution Development

The following sections provide the research findings according to the main Design Science Research phases of Hevner and Chatterjee [29].

4.1. Suggestion Phase: Verba RAG

For the development of an RAG system and to address the identified problems (as stated in Section 4.1), Verba was used. Verba is an open-source RAG application that leverages Weaviate, a vector database, for information retrieval. Using the Weaviate Cloud Service (WCS), one can easily implement Verba RAG without extensive technical knowledge [39].

4.1.1. Verba RAG Architecture

Verba's modular architecture enables the customization of various components (see Table 3), such as readers, chunkers, embedders, retrievers, and generators [39], as illustrated in Figure 2.

Table 3. Verba RAG architecture.

Name	Task
Read-Chunk Manager	The component (https://weaviate.io/blog/verba-open-source-rag-app#chunker-manager-breaking-data-up, accessed on 10 June 2024) receives a list of strings representing uploaded documents. In our case, it handles the reports collected from various banks. The Read-Chunk Manager takes a list of documents and breaks each document's text into smaller segments. For this use case, the Read-Chunk Manager divides each document into chunks of 100-word tokens with a 50-token overlap. This method will remain consistent across all the tests. This approach is based on preliminary testing, which clearly indicated that larger chunk sizes (e.g., 250- and 400-word tokens) negatively affected the quality of the final output of the proposed RAG system when applied to the financial reports.

Table 3. *Cont.*

Name	Task
Embedding Manager	The Embedding Manager (https://weaviate.io/blog/verba-open-source-rag-app#embedding-manager-vectorizing-data, accessed on 10 June 2024) receives a list of documents and embeds them as vectors into Weaviate as the relevant database. It is also used to retrieve chunks and documents from Weaviate. The specific embedding model used will vary according to the model configuration described in Section 4.2.1.
Retrieve Manager	The Retrieve Manager (https://weaviate.io/blog/verba-open-source-rag-app#retrieve-manager-finding-the-context, accessed on 10 June 2024) communicates with the Embedding Manager to retrieve chunks and apply custom logic. They return a list of chunks. For this use case, the "WindowRetriever" is employed for all the tests, which retrieves relevant chunks and their surrounding context using a combination of semantic and keyword search (hybrid approach).
Generation Manager	The Generation Manager (https://weaviate.io/blog/verba-open-source-rag-app#generation-manager-writing-the-answer, accessed on 10 June 2024) uses a list of chunks and a query to generate an answer. Then, it returns a string as the answer. Like the Embedding Manager, the specific generating model used will vary according to the model configuration described in Section 4.2.1.

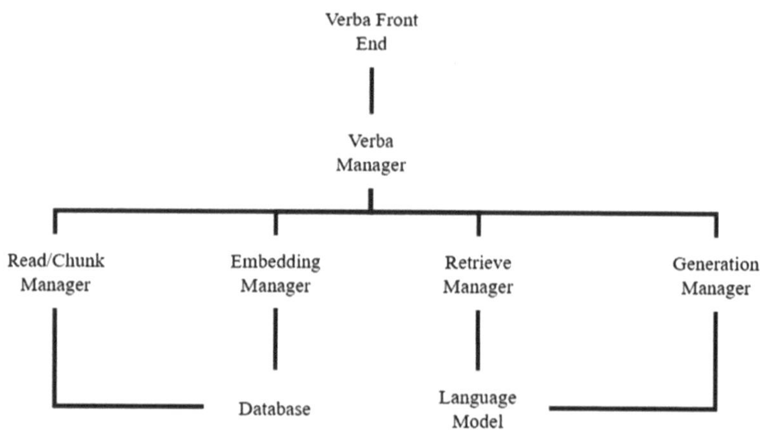

Figure 2. Illustration of the Verba RAG architecture based on [39].

4.1.2. Verba Flow

Verba by Weaviate [39] is designed to be customizable and scalable. The flow of this RAG system is generally the same as described in Section 3.1.3: Users can upload or import data files via the Verba web interface. Once uploaded, Verba automatically chunks the documents and stores the embeddings in the WCS Weaviate instance. When a query is entered into the Verba web interface, relevant document chunks are retrieved from the WCS Weaviate instance based on the selected query embeddings. These retrieved chunks are then passed to a language model (e.g., GPT-4o) to generate a contextual answer. Verba displays the generated answer along with the relevant document chunks used for context.

4.2. Development Phase: Experiments with Verba RAG

As detailed in Section 3.1.2, the artifact for this project involves the construction of an RAG pipeline with Verba by Weaviate and various other components. To address research questions one and two of Section 1.2, the RAG system will be developed, and three model configurations will be created, each using different components. The setup and design of the experiments are outlined in the following sections.

4.2.1. Model Configurations

Three model configurations with different components were developed for the evaluation of the RAG system, as shown in Table 4.

Table 4. Model configurations.

Model	Embedding Model (Part of "Retrieval")	LLM (Part of "Generation")	Purpose
Baseline Model	OpenAI ADA	OpenAI GPT-4o	Establish a benchmark for performance based on widely used, reliable components.
Second Model	OpenAI ADA (consistent with baseline)	Google Gemini 1.5 Pro	Evaluate the impact of substituting the LLM (generator component) with a newly released LLM [1] while keeping the embedding model constant to isolate the effect.
Third Model	MiniLM Embedder	OpenAI GPT-4o (consistent with baseline)	Assess the effect on the overall performance of changing the embedding model (part of the retrieval component) to an open-source component while using the same LLM.

[1] according to the official website of OpenAI [41], GPT-4o was launched on 13 May 2024, whereas Google [42,43] released Gemini 1.5 Pro on 9 April 2024 to the public and upgraded it on May 14, 2024. The latest models are used.

4.2.2. Data

For each model configuration, two quarterly and two half-yearly reports from the five selected banks are used, as described in Section 3.2. To introduce complexity and simulate real-world challenges, reports from three additional banks are included.

4.2.3. Questions

Each bank will be subjected to ten specific questions and three model configurations. Seven of these questions are general, applicable across all the banks, and designed to evaluate the system's general capabilities. These general questions are a mix of quantitative (e.g., financial metrics like revenue or profit) and qualitative types (e.g., risks, challenges, and trends), but all are questions that a private investor could ask. They were developed iteratively, starting with simple queries and refining them through experiments to determine the optimal length and style. We ended up with relatively long questions that accommodate different wordings, such as revenue, total income, and gross earnings, as banks often use these terms synonymously. These questions can be found in Appendix A.

The remaining three questions are tailored to each individual bank to assess the system's ability to handle specific and detailed inquiries.

4.2.4. Experiments

The experiments were conducted by providing the RAG system with ten questions for each bank and model configuration. To ensure precise analysis and response, these questions were submitted individually in a single session rather than in batches. Each question, answer, and the retrieved context were compiled into an Excel. The complete file includes data for five banks for three model configurations with ten questions per model configuration.

The RAG system was set up according to the identified requirements, developing three distinct model configurations with different components. This setup addresses research questions 1 and 2 from Section 1.2, particularly defining the retrieval and generation components. On the other hand, the defined and developed setup of the RAG system and the experiments form the basis for answering research questions three and four.

5. Evaluation and Discussion of Results

This section outlines how the evaluation of the experiments was conducted. The results of the evaluation are then discussed. This section addresses research questions 3 and 4 of Section 1.2. There is also a discussion of the academic relevance of this study at the end of this section.

5.1. Realization of Evaluation

To evaluate the three model configurations and assess the RAG systems, three quality metrics were used: context relevance (retrieval metric), answer faithfulness (generation metric), and answer relevance (generation metric) (described in Section 2.3).

Each quality metric is evaluated manually for each response, corresponding to a specific question for a bank and model configuration (as outlined in Section 3.1.3). Each metric is rated on a scale from 1 to 5 specifically developed for this purpose (see Table 5).

Table 5. Scale for evaluation.

Scale	Definition
1	The retrieved context is irrelevant to the question, the generated response is inaccurate and inconsistent, and it fails to address the question.
2	The retrieved context has limited relevance and contains several inaccuracies, the generated response exhibits notable inconsistencies, and it only partially addresses the given question.
3	The retrieved context is somewhat relevant but has some inaccuracies, the generated response is generally consistent with minor errors, and it adequately addresses the question.
4	The retrieved context is relevant and mostly accurate, the generated response is consistent with minor or no errors, and it effectively addresses the question.
5	The retrieved context is highly relevant and accurate, the generated response is entirely faithful to the context with no errors, and it thoroughly addresses the question.

A total of 750 points are possible for each model configuration, as there are five banks with 10 questions for each model, with five points possible per question in three different categories. The maximum number of points per model configuration is calculated as follows: $5 \times 10 \times 5 \times 3 = 750$ points.

5.2. Main Results and Discussion

The primary outcome is the evaluation of the three model configurations (see Table 6). Our initial goal was to determine which model configuration performed best and to address research questions 3 and 4 from Section 1.2.

Table 6. Evaluation results: three model configurations (best results are indicated in bold).

No.	Embedding Model	LLM	Points	Metric 1: Context Relevance	Metric 2: Answer Faithfulness	Metric 3: Answer Relevance	Total
1	OpenAI ADA	OpenAI GPT-4o	Total Points	175	**189**	**188**	**552**
			Average Points	**3.5**	**3.78**	**3.76**	**3.68**
2	OpenAI ADA (consistent with baseline)	Google Gemini 1.5 Pro	Total Points	**176**	185	182	543
			Average Points	**3.52**	3.7	3.64	3.62
3	MiniLM Embedder	OpenAI GPT-4o (consistent with baseline)	Total Points	137	155	146	438
			Average Points	2.74	3.1	2.92	2.92

Model one (OpenAI ADA and OpenAI GPT-4o) achieved the highest overall score (552 of 750 points). With an average score of 3.7 out of 5, this model configuration falls between the 3- and 4-point definitions on the scale presented in Section 5.1. Thus, the retrieved context was generally relevant and mostly accurate, and the generated responses were consistent with minor errors to address the posed questions. The lowest scores were assigned to "Context Relevance", often due to retrieving over 10–15 chunks, many of which were not pertinent. However, even with irrelevant chunks, the LLM still provided mostly accurate answers. This highlights GPT-4o's robustness in handling excess information.

Model two (OpenAI ADA and Gemini 1.5 Pro) received a slightly lower overall score. The "Context Relevance" score remained nearly stable due to using the same embedding model. However, the scores for "Answer Faithfulness" and "Answer Relevance" were slightly lower, reflecting the impact of using a different LLM. This observation aligns with benchmarks from the LLM Leaderboard of the Large Model Systems Organization [44], where GPT-4o was rated with 1287 points compared to Gemini 1.5 Pro with 1266 points. In addition, the Massive Multitask Language Understanding (MMLU) benchmark developed by Hendrycks et al. [45] obtained a score of 88.7% for GPT-4o and 81.9% for Gemini 1.5 Pro [46]. These benchmarks indicate a slight performance edge for GPT-4o, which was reflected in the evaluation results.

Model three (MiniLM Embedder and OpenAI GPT-4) demonstrated a significantly lower overall score of 438 points, which was more than 20% lower than model one and lower than model two. Changing the embedding model from OpenAI's ADA to the MiniLM Embedder resulted in lower scores for all the metrics, which is not unexpected as the new model is much smaller than ADA. The "Context Relevance" score was considerably lower because much of the retrieved context was not relevant to the posed questions. We observed that this model configuration often retrieved fewer chunks than the previous models, which also affected "Answer Faithfulness" and "Answer Relevance". Without sufficient (quantity) and relevant (quality) context, the LLM struggled to generate accurate answers. This underscores the importance of using a high-quality embedding model to retrieve relevant information.

In summary, to answer research questions three and four from Section 1.2, models one and two demonstrated a more accurate representation of the necessary information, while model three lagged behind. The evaluation approach and setup were effective for reliably evaluating and validating these experiments.

5.3. Additional Results and Discussion

Table 7 presents the evaluation results for each bank, independent of the model configuration. The main difference was observed between Barclays (290 of 450 points) and HSBC (338 of 450 points). Barclays, with an average score of 3.2, had a notably low score for "Context Relevance". A comparison of the half-yearly and quarterly reports of Barclays and HSBC reveals that they use completely different layouts. Barclays reports are in landscape mode and contain a large amount of data displayed in tables and various types of charts (e.g., line charts and bar charts) with minimal written text. On the contrary, HSBC's reports are text-heavy with well-explained tables and minimal charts. This layout makes it easier for retrievers to gather the relevant context. Consequently, the HSBC reports are more suitable for the RAG system.

Table 7. Results of evaluation—total points per bank (best results are indicated in bold).

Bank	Metric 1: Context Relevance	Metric 2: Answer Faithfulness	Metric 3: Answer Relevance	Total
Banco Santander	96.0	97.0	99.0	292.0
CS	104.0	97.0	97.0	298.0
Barclays	81.0	107.0	102.0	290.0
Credit Agricole	**106.0**	107.0	102.0	315.0
HSBC	101.0	**121.0**	**116.0**	**338.0**

Table 8 presents the results for each of the seven standards and three individual questions. The highest scores were obtained on individual questions. This indicates that when specific questions and wording align with the report, the RAG system performs well across all three metrics, thereby providing accurate answers.

Table 8. Results of evaluation—total points per question (best results are indicated in bold).

Questions	Metric 1: Context Relevance	Metric 2: Answer Faithfulness	Metric 3: Answer Relevance	Total
Indiv. Question 1	57	58	59	174
Indiv. Question 2	54	59	58	171
Indiv. Question 3	**63**	**63**	**62**	**188**
Standard Question 1 (quantitative)	42	56	51	149
Standard Question 2 (quantitative)	45	49	45	139
Standard Question 3 (quantitative)	45	46	45	136
Standard Question 4 (quantitative)	42	46	44	132
Standard Question 5 (qualitative)	43	46	46	135
Standard Question 6 (qualitative)	47	54	53	154
Standard Question 7 (qualitative)	50	52	53	155

The qualitative questions also received higher average scores than the quantitative questions. It was found that when multiple similar numbers appeared in a report, the RAG sometimes had difficulties handling them accurately. For example, question 4 asked for the total assets held by the bank at the end of H1 2023. Banks typically report various types of assets, such as risk-weighted assets, customer assets, or high-quality liquid assets. If the RAG system selected the wrong number, it received a low score. However, minimal hallucination was observed; instead, the primary issue was selecting an incorrect but actually appearing number. On the other hand, for qualitative questions such as market trends, the RAG system often produced at least a partially correct answer, which resulted in higher scores.

5.4. Academic Relevance of Artifact Evaluation

The evaluation highlights the academic significance of the RAG system in assisting private investors with bank financial report analysis (see Section 2.4). By leveraging RAG, investors can access domain-specific information from half-yearly or quarterly reports. Among the evaluated model configurations, model one was the top performer, delivering accurate answers with minor errors. However, due to the variability in performance, the importance of selecting reliable system components was underscored.

This study shows the critical role of well-structured financial reports, as poorly organized documents, such as those from Barclays, received lower scores. This highlights the necessity for clear and coherent financial documentation. In addition, this study identifies the types of questions that the RAG system handles most proficiently, providing valuable academic insights.

Importantly, the evaluation has shown that the RAG concept addresses challenges in LLMs, particularly issues related to factual accuracy and the avoidance of hallucinations. These findings underscore the crucial importance of this research area for both academic inquiry and practical applications in financial report analysis.

6. Conclusions

This section consolidates our research findings, assesses the success of addressing the main research question, and presents the key insights gained from the study. Practical implications, limitations, and the potential future developments of the suggested artifact are discussed.

The study focused on the main research question: "How can the RAG be adapted to increase the context relevance, answer faithfulness, and answer relevance of LLM's conclusions about several types of financial reports?". By developing an RAG system and exploring three technical RAG model configurations employing different components (embedding models and LLMs), variations in these metrics were identified. The development and evaluation process led us to conclude that the first model configuration was particularly effective. Ultimately, the findings suggest that we can empower private investors to make informed decisions when provided with accurate answers derived by such an RAG system from relevant bank reports.

Although the main research question was addressed successfully, the study revealed several areas for improvement. A significant limitation was the RAG system's difficulty in processing complex PDF layouts. Future research should aim to optimize the system by integrating more robust components, such as computer vision models, to handle visual information more effectively. Moreover, adding a domain-specific repository for financial terminology would enhance context understanding and mitigate issues arising from different terminology used by different banks for similar concepts (e.g., assets).

Future research holds various promising avenues to provide investors with even more reliable insights. Expanding the system to include a broader spectrum of banks is a critical aspect. In addition, exploring more advanced evaluation methods could enhance the robustness and accuracy of the assessment process, thereby improving RAG's overall efficacy.

By addressing these limitations and exploring the outlined research directions, the RAG systems' capability to analyze financial reports and deliver valuable insights to private investors could be enhanced.

Author Contributions: Methodology, software, investigation, and writing—original draft preparation, I.I., R.P., and L.V.; writing—review and editing and supervision, T.H. All authors have read and agreed to the published version of the manuscript.

Funding: This research received no external funding.

Data Availability Statement: The data presented in this study are available upon request from the corresponding author due to legal reasons.

Conflicts of Interest: The authors declare no conflicts of interest.

Appendix A

Table A1. Questions for Experiments.

No.	Question/Prompt	Period	Type
1.	What was the total revenue for [Bank Name] in quarter Q1 2022? Retrieve the total revenue for [Bank Name] from the report of the quarter Q1 2022. Also, consider alternative terms such as 'total income' or 'gross earnings' if they are used in the document.	Quarter	Quantitative
2.	What is the percentage change in total revenue for [Bank Name] in Q1 2023 compared to the same quarter of the previous year? Find or calculate the percentage change in total revenue for [Bank Name] of Q1 2023 compared to the to the same quarter of the previous year. Include terms like 'percentage growth' or 'percentage decline' if mentioned.	Quarter	Quantitative
3.	What is the net profit reported by [Bank Name] for the half-year period H1 2022? Identify the net profit for [Bank Name] for the half-year period H1 2022, including any references to 'net income' or 'net earnings'. Ensure the figure reflects income after all expenses.	Half-yearly	Quantitative
4.	What are the total assets held by [Bank Name] at the end of the half-yearly period H1 2023? Extract the total assets held by [Bank Name] at the end of the half-yearly period H1 2023. Include all categories such as cash, investments, loans, and other financial instruments.	Half-yearly	Quantitative
5.	What strategic initiatives or projects did [Bank Name] undertake during the reporting period Q1 2022? Summarize the strategic initiatives or projects undertaken by [Bank Name] during the reporting period, including new product launches, market expansions, and digital transformation efforts. There might be also similar strategic initiatives or projects.	Quarter	Qualitative
6.	What trends or market conditions affecting the banking sector were highlighted by [Bank Name] in their report of the half-yearly period H1 2022? Highlight the trends or market conditions affecting the banking sector as discussed by [Bank Name], such as interest rate changes, regulatory impacts, and technological advancements. There might be also similar trends and market conditions.	Half-yearly	Qualitative
7.	What key risks and challenges for the upcoming period did [Bank Name] identify in their report of the half-yearly period H1 2023?? Detail the key risks and challenges for the upcoming period identified by [Bank Name] include, including any mention of credit risk, market volatility, and operational challenges. There might be also similar risks and key challenges.	Half-yearly	Qualitative

References

1. Zhao, W.X.; Zhou, K.; Li, J.; Tang, T.; Wang, X.; Hou, Y.; Min, Y.; Zhang, B.; Zhang, J.; Dong, Z.; et al. A Survey of Large Language Models. *arXiv* **2023**, arXiv:2303.18223.
2. Brown, T.; Mann, B.; Ryder, N.; Subbiah, M.; Kaplan, J.D.; Dhariwal, P.; Neelakantan, A.; Shyam, P.; Sastry, G.; Askell, A. Language Models are Few-Shot Learners. *Adv. Neural Inf. Process. Syst.* **2020**, *33*, 1877–1901.
3. Burtsev, M.; Reeves, M.; Job, A. The Working Limitations of Large Language Models. *MIT Sloan Manag. Rev.* **2024**, *65*, 8–10.
4. Guu, K.; Lee, K.; Tung, Z.; Pasupat, P.; Chang, M. REALM: Retrieval-Augmented Language Model Pre-Training. In Proceedings of the International Conference on Machine Learning, PMLR, Vienna, Austria, 13–18 July 2020; pp. 3929–3938.
5. Lewis, P.; Perez, E.; Piktus, A.; Petroni, F.; Karpukhin, V.; Goyal, N.; Küttler, H.; Lewis, M.; Yih, W.; Rocktäschel, T. Retrieval-Augmented Generation for Knowledge-Intensive NLP Tasks. *Adv. Neural Inf. Process. Syst.* **2020**, *33*, 9459–9474.
6. Gao, Y.; Xiong, Y.; Gao, X.; Jia, K.; Pan, J.; Bi, Y.; Dai, Y.; Sun, J.; Guo, Q.; Wang, M.; et al. Retrieval-Augmented Generation for Large Language Models: A Survey. *arXiv* **2024**, arXiv:2312.10997.
7. IBM Research. What Is Retrieval-Augmented Generation? Available online: https://research.ibm.com/blog/retrieval-augmented-generation-RAG (accessed on 9 February 2021).
8. Chen, J.; Lin, H.; Han, X.; Sun, L. Benchmarking large language models in retrieval-augmented generation. In Proceedings of the AAAI Conference on Artificial Intelligence, Vancouver, BC, Canada, 20–27 February 2024; Volume 38, pp. 17754–17762.
9. Shuster, K.; Poff, S.; Chen, M.; Kiela, D.; Weston, J. Retrieval Augmentation Reduces Hallucination in Conversation. *arXiv* **2021**, arXiv:2104.07567.
10. Anil, R.; Borgeaud, S.; Wu, Y.; Alayrac, J.-B.; Yu, J.; Soricut, R.; Schalkwyk, J.; Dai, A.M.; Hauth, A.; Millican, K.; et al. Gemini: A family of highly capable multimodal models. *arXiv* **2023**, arXiv:2312.11805.
11. Touvron, H.; Martin, L.; Stone, K.; Albert, P.; Almahairi, A.; Babaei, Y.; Bashlykov, N.; Batra, S.; Bhargava, P.; Bhosale, S.; et al. Llama 2: Open Foundation and Fine-Tuned Chat Models. *arXiv* **2023**, arXiv:2307.09288.
12. He, H.; Zhang, H.; Roth, D. Rethinking with Retrieval: Faithful Large Language Model Inference. *arXiv* **2022**, arXiv:2301.00303.
13. Shen, X.; Chen, Z.; Backes, M.; Zhang, Y. In ChatGPT We Trust? Measuring and Characterizing the Reliability of ChatGPT. *arXiv* **2023**, arXiv:2304.08979.
14. Zhang, Y.; Li, Y.; Cui, L.; Cai, D.; Liu, L.; Fu, T.; Huang, X.; Zhao, E.; Zhang, Y.; Chen, Y.; et al. Siren's Song in the AI Ocean: A Survey on Hallucination in Large Language Models. *arXiv* **2023**, arXiv:2309.01219.
15. Borgeaud, S.; Mensch, A.; Hoffmann, J.; Cai, T.; Rutherford, E.; Millican, K.; Van Den Driessche, G.B.; Lespiau, J.-B.; Damoc, B.; Clark, A. Improving language models by retrieving from trillions of tokens. In Proceedings of the International Conference on Machine Learning, PMLR, Baltimore, MD, USA, 17–23 July 2022; pp. 2206–2240.
16. Lyu, Y.; Li, Z.; Niu, S.; Xiong, F.; Tang, B.; Wang, W.; Wu, H.; Liu, H.; Xu, T.; Chen, E.; et al. CRUD-RAG: A Comprehensive Chinese Benchmark for Retrieval-Augmented Generation of Large Language Models. *arXiv* **2024**, arXiv:2401.17043.
17. Es, S.; James, J.; Espinosa-Anke, L.; Schockaert, S. RAGAS: Automated Evaluation of Retrieval Augmented Generation. *arXiv* **2023**, arXiv:2309.15217.
18. Saad-Falcon, J.; Khattab, O.; Potts, C.; Zaharia, M. ARES: An Automated Evaluation Framework for Retrieval-Augmented Generation Systems. *arXiv* **2023**, arXiv:2311.09476.
19. Ragas. Core Concepts. 2023. Available online: https://docs.ragas.io/en/latest/concepts/index.html (accessed on 10 July 2024).
20. TruLens. RAG Triad. 2024. Available online: https://truera.com/ai-quality-education/generative-ai-rags/what-is-the-rag-triad/ (accessed on 10 July 2024).
21. Besbes, A. A 3-Step Approach to Evaluate a Retrieval Augmented Generation (RAG). Towards Data Science. Available online: https://towardsdatascience.com/a-3-step-approach-to-evaluate-a-retrieval-augmented-generation-rag-5acf2aba86de (accessed on 23 November 2023).
22. Besbes, A. Quickly Evaluate Your RAG without Manually Labeling Test Data. Towards Data Science. Available online: https://towardsdatascience.com/quickly-evaluate-your-rag-without-manually-labeling-test-data-43ade0ae187a (accessed on 21 December 2023).
23. Frenchi, C. Evaluating RAG: Using LLMs to Automate Benchmarking of Retrieval Augmented Generation Systems. Willow Tree Apps. Available online: https://www.willowtreeapps.com/craft/evaluating-rag-using-llms-to-automate-benchmarking-of-retrieval-augmented-generation-systems (accessed on 1 December 2023).
24. Leal, M.; Frenchi, C. Evaluating Truthfulness: Benchmarking LLM Accuracy. Willow Tree Apps. Available online: https://www.willowtreeapps.com/craft/evaluating-truthfulness-a-deeper-dive-into-benchmarking-llm-accuracy (accessed on 21 September 2023).
25. Nguyen, R. LlamaIndex: How to Evaluate Your RAG (Retrieval Augmented Generation) Applications. Better Programming. Available online: https://betterprogramming.pub/llamaindex-how-to-evaluate-your-rag-retrieval-augmented-generation-applications-2c83490f489 (accessed on 1 October 2023).
26. Sarmah, B.; Zhu, T.; Mehta, D.; Pasquali, S. Towards reducing hallucination in extracting information from financial reports using Large Language Models. *arXiv* **2023**, arXiv:2310.10760.
27. Yepes, A.J.; You, Y.; Milczek, J.; Laverde, S.; Li, R. Financial Report Chunking for Effective Retrieval Augmented Generation. *arXiv* **2024**, arXiv:2402.05131.

28. Zhang, B.; Yang, H.; Zhou, T.; Ali Babar, M.; Liu, X.-Y. Enhancing Financial Sentiment Analysis via Retrieval Augmented Large Language Models. In Proceedings of the Fourth ACM International Conference on AI in Finance, Brooklyn, NY, USA, 27–29 November 2023; pp. 349–356.
29. Hevner, A.; Chatterjee, S. (Eds.) Design Science Research in Information Systems. In *Design Research in Information Systems: Theory and Practice*; Springer: Berlin/Heidelberg, Germany, 2010; pp. 9–22. [CrossRef]
30. HuggingFace. MiniLMEmbedder. Available online: https://huggingface.co/sentence-transformers/all-MiniLM-L6-v2 (accessed on 18 January 2024).
31. LangChain. Bedrock Embeddings. 2023. Available online: https://python.langchain.com/v0.1/docs/integrations/text_embedding/bedrock/ (accessed on 10 July 2024).
32. Meta. Faiss: A Library for Efficient Similarity Search. Available online: https://engineering.fb.com/2017/03/29/data-infrastructure/faiss-a-library-for-efficient-similarity-search/ (accessed on 29 March 2017).
33. Chroma. Chroma Docs. 2024. Available online: https://docs.trychroma.com/getting-started (accessed on 10 July 2024).
34. OpenAI. OpenAI Platform. 2024. Available online: https://platform.openai.com/docs/overview (accessed on 10 July 2024).
35. Meta. Meta Llama 3. 2024. Available online: https://llama.meta.com/llama3/ (accessed on 10 July 2024).
36. Google DeepMind. Gemini Pro 1.5. Available online: https://deepmind.google/technologies/gemini/pro/ (accessed on 20 May 2024).
37. LangChain. Introduction to LangChain. 2023. Available online: https://python.langchain.com/v0.1/docs/get_started/introduction/ (accessed on 10 July 2024).
38. LlamaIndex. LlamaIndex Docs. 2024. Available online: https://docs.llamaindex.ai/en/stable/ (accessed on 10 July 2024).
39. Weaviate. Verba Docs. GitHub. 2024. Available online: https://github.com/weaviate/Verba (accessed on 10 July 2024).
40. Weaviate. Verba—Demo Tool. 2024. Available online: https://verba.weaviate.io/ (accessed on 10 July 2024).
41. OpenAI. Hello GPT-4o. Available online: https://openai.com/index/hello-gpt-4o/ (accessed on 13 May 2024).
42. Google. Gemini 1.5 Pro Now Available in 180+ Countries. Available online: https://developers.googleblog.com/en/gemini-15-pro-now-available-in-180-countries-with-native-audio-understanding-system-instructions-json-mode-and-more/ (accessed on 9 April 2024).
43. Google. Gemini 1.5 Pro Updates. Available online: https://blog.google/technology/developers/gemini-gemma-developer-updates-may-2024/ (accessed on 14 May 2024).
44. Large Model Systems Organization. LLM Leaderboard. Available online: https://chat.lmsys.org/?leaderboard (accessed on 6 June 2024).
45. Hendrycks, D.; Burns, C.; Basart, S.; Zou, A.; Mazeika, M.; Song, D.; Steinhardt, J. Measuring Massive Multitask Language Understanding. In Proceedings of the 9th International Conference on Learning Representations, ICLR 2021, Virtual Event, Austria, 3–7 May 2021; pp. 11260–11285.
46. Papers with Code. Multi-task Language Understanding on MMLU. 2024. Available online: https://paperswithcode.com/sota/multi-task-language-understanding-on-mmlu (accessed on 10 July 2024).

applied sciences

MDPI

Article

A Graph Convolutional Network Based on Sentiment Support for Aspect-Level Sentiment Analysis

Ruiding Gao [1], Lei Jiang [1,*], Ziwei Zou [1], Yuan Li [1] and Yurong Hu [2]

[1] School of Computer Science and Engineering, Hunan University of Science and Technology, Xiangtan 411201, China; 2105060119@mail.hnust.edu.cn (R.G.); 20010501001@mail.hnust.edu.cn (Z.Z.); 20010501008@mail.hnust.edu.cn (Y.L.)

[2] School of Computer Engineering, Jingchu University of Technology, Jingmen 448000, China; yrhu@jcut.edu.cn

* Correspondence: jleihn@hnust.edu.cn

Abstract: Aspect-level sentiment analysis is a research focal point for natural language comprehension. An attention mechanism is a very important approach for aspect-level sentiment analysis, but it only fuses sentences from a semantic perspective and ignores grammatical information in the sentences. Graph convolutional networks (GCNs) are a better method for processing syntactic information; however, they still face problems in effectively combining semantic and syntactic information. This paper presents a sentiment-supported graph convolutional network (SSGCN). This SSGCN first obtains the semantic information of the text through aspect-aware attention and self-attention; then, a grammar mask matrix and a GCN are applied to preliminarily combine semantic information with grammatical information. Afterward, the processing of these information features is divided into three steps. To begin with, features related to the semantics and grammatical features of aspect words are extracted. The second step obtains the enhanced features of the semantic and grammatical information through sentiment support words. Finally, it concatenates the two features, thus enhancing the effectiveness of the attention mechanism formed from the combination of semantic and grammatical information. The experimental results show that compared with benchmark models, the SSGCN had an improved accuracy of 6.33–0.5%. In macro F1 evaluation, its improvement range was 11.68–0.5%.

Keywords: aspect-level sentiment analysis; graph convolutional network; attention mechanisms; sentiment support words

1. Introduction

Sentiment analysis involves extracting viewpoints and judging emotional tendencies in subjective texts. Aspect-level sentiment analysis is more refined, particularly when judging the emotional polarity of texts in terms of given aspect words (e.g., positive, negative, and neutral) [1–3]. As shown in Figure 1, the sentence has two sides: "price" and "screen". The emotional polarity of the sentence is negative with regard to "price" and positive in terms of "screen". The main task of aspect-level sentiment analysis is to accurately find the opinions and emotional tendencies of sentences according to the identified aspects. In this way, the statement is partially understood. This is currently an important tool for online opinion gathering for governments and businesses concerning policies or business strategies [4]. For example, it can be used to analyze a company's blockchain statements and obtain a short-term value evaluation of the company [5]. It can also be employed to analyze the behavior of users in the blockchain [6].

Attention mechanisms can improve the structure of deep learning by concentrating on the important points. Wang et al. [7] presented the application of attention mechanisms to long short-term memory (LSTM) [8]. They used an attention mechanism to acquire the semantic information between words in a text and aspect words, obtaining crucial parts of the sentences to obtain a more accurate aspect sentiment analysis. Ma et al. [9] introduced interactive attention networks (IANs) based on LSTM and attention mechanisms to realize the interactive learning of attention based on goals and contexts. However, many studies have used attention mechanisms to obtain the relation between words and aspect words in sentences from a semantic perspective [10–12]. When a sentence has several aspect words and the emotional polarity of these words is different, the attention mechanism can easily focus on the wrong part or a part that is unrelated to the aspect words. As shown in Figure 1, if the attention mechanism mistakenly concentrates on the opinion word "good" for the aspect of "price", it is likely to cause errors in the emotional polarity.

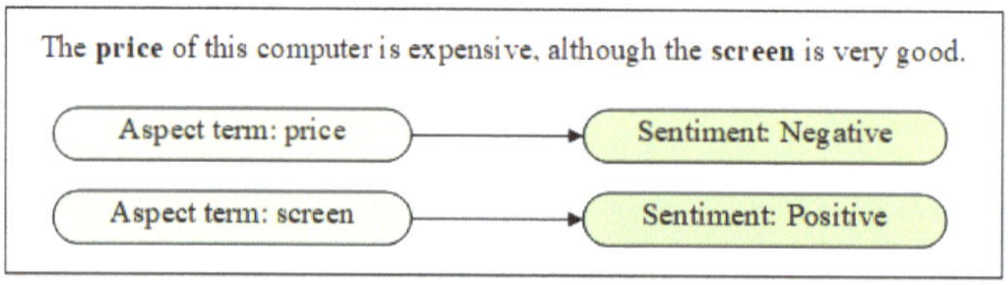

Figure 1. An example sentence of aspect-level sentiment analysis.

In order to enable the model to accurately concentrate on opinion words that correspond to the aspect words, it is important to introduce the grammatical relations between words in sentiment analysis. Graph convolutional networks (GCNs) [12] combined with syntactic dependency trees can better obtain grammatical features. Zhang C. et al. [10] employed graph convolutional networks in aspect-level sentiment analysis, and their results showed excellent model performance. In the following study, Zhang Z. et al. [11] joined the grammatical structure and semantic correlation of aspect words. They presented an aspect perception attention mechanism combined with a self-attention mechanism, and their work shows promising results.

It is evident that a sentence not only contains aspect words but also other types of words. If sentiment analysis was only performed with the combination of grammatical and semantic features of aspect words, some important information in the sentence would be lost. Therefore, we introduced sentiment support words (SSWs; proposed in our previous study [13]) to supplement a sentiment analysis in order to obtain more complete grammatical and semantic features of sentences. SSWs are words that influence an aspect word in a sentence, often supporting the emotional polarity of the whole sentence in one aspect. Through a grammatical dependency tree, we can obtain the aspect of sentiment support words. Figure 2 shows a concrete example of this. From the figure, we can see that the aspect words are attributes of the object entities in the sentence. For example, an aspect can be the price of computers or computer screens, while sentiment support words directly evaluate part of the target information. For example, when the aspect is the price of computers, sentiment support words evaluate the price of computers as "expensive". The emotional polarity here is clearly negative. Therefore, combining aspect words with SSWs is beneficial for identifying the emotional polarity corresponding to certain aspect words. We can identify the SSWs corresponding to each aspect of a sentence by analyzing the parts of speech and applying fixed rules. This approach allows us to consider the grammatical and semantic information of emotive support words in the sentence.

Based on this idea, this paper presents a sentiment-supported graph conventional network (SSGCN). First, we used an aspect perception attention mechanism and a self-attention mechanism to obtain attention scores; then, we input these scores into the GCN through a syntactic mask matrix to obtain sentence features that initially combined semantic and grammatical information. Then, the semantic and grammatical information of the SSW combination was obtained. Finally, this was combined with the semantic and grammatical information of the aspect words to obtain the enhanced information of the combination of syntactic and semantic information. In this way, a more complete combination of semantic and grammatical information could be obtained. When evaluated on three publicly available datasets, the model presented in this paper demonstrated a greater efficiency than the eight comparison models.

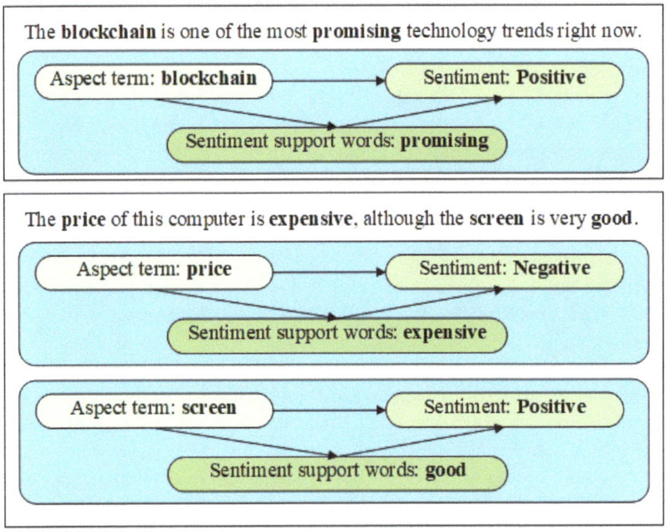

Figure 2. Examples of SSWs in aspect-level sentiment analysis sentences.

The main contributions of this paper are as follows:

(1) A method of combining semantics and grammar using sentiment support words is presented. This method effectively captures the evaluation information of aspect words in sentences and assists the model in extracting the correct feature information.
(2) A further combination method is presented. This method integrates syntactic and semantic information based on aspect and sentiment support words. It enhances the comprehensiveness of the constructed combined features by incorporating semantic and grammatical information.
(3) Experiments on three public datasets demonstrated that the SSGCN was more effective than the benchmark models.

The remainder of this paper is organized as follows: Section 2 provides a review of the research associated with aspect-level sentiment analysis. Then, the proposed SSGCN model is introduced in detail in Section 3. The experiments and discussion are presented in Section 4. Finally, Section 5 provides our conclusions.

2. Related Work

Aspect-level sentiment analysis is a subtask of sentiment analysis. Compared with ordinary sentiment analysis tasks, it is a more refined emotional classification task aimed at discovering the emotional polarity of a sentence in a specific aspect.

Early sentiment analysis tasks mostly adopted machine learning methods [14]. Kiritchenko et al. [15] obtained a large number of analytical attributions for classification by manually constructing features. However, this method is labor-intensive, demands a lot of human resources and time, and loses the dependency between the aspect words and the context. Kim et al. [16] used neural networks to model text, concluding that convolutional neural networks (CNNs) perform well when extracting the local features of sentences in a text. Therefore, CNNs have begun to be widely used in sentiment analysis [17–19]. The variant model of the recurrent neural network (RNN) is also a common method used in sentiment analysis [20]. For instance, in the case of using aspect word information, Tang et al. [21] introduced target-dependent long short-term memory (TD-LSTM). This approach employs two LSTMs to model the context before and after aspect words, utilizing the aspect words' semantic information to judge the emotional polarity. This method achieved good results in their experiments. In aspect-level sentiment analysis, an attention-mechanism-based approach [22,23] directs the model's increased attention toward the feature information of aspect words. For instance, Wang et al. [7] input the word vector of aspect words into the structure of an attention mechanism. The obtained attention score represents the weight of each word in this aspect of the sentence. The higher the score, the more important it becomes. Finally, the sentence features are obtained by weighted summation. Their experiments showed that this method is superior to its baseline methods. Similarly, Huang et al. [24] used BiLSTM-SNP to obtain the correlation between aspect words and content words, applying attention mechanisms to determine their weights. The experiments on multiple Chinese and English datasets showed that this method has significant advantages in terms of accuracy.

To a certain extent, the attention mechanism can find the semantic correlation between aspect words and other words in a sentence, but when the sentence composition is complex, the attention mechanism may make the model concentrate on the wrong part of the sentence [25]. Unlike traditional deep learning models, graph convolutional networks capture the relationships and feature information between nodes in a graph by defining the convolution operations on the graph structure [26]. Specifically, they first represent nodes in the graph as low-dimensional vectors and aggregate the information of adjacent nodes through matrix multiplication to obtain new feature representations for each node in the graph. Since GCNs address this deficiency better [27], it was used for the first time by Zhang et al. [10] for aspect-level sentiment analysis. Their proposed method was named the aspect-specific graph convolutional network (ASGCN). The model established the GCN on the syntactic dependency tree and acquired grammar-dependent information from a sentence. The dual graph convolutional network (DualGCN) [28] was proposed by Li et al. They also argued for the complementarity of grammatical structure and semantic relations, and combined an attention mechanism with a grammatical dependency tree. In this model, a GCN is employed to obtain the abundant syntactic information on the syntactic dependency tree, and another GCN is applied to obtain the semantic relations using a self-attention mechanism. Yang et al. [29] introduced tight connections into GCNs, enabling GCNs to obtain rich local and global information. A syntactic- and semantic-enhanced graph convolutional network (SSEGCN) was presented by Zhang et al. [11]. It initially integrates the syntactic and semantic information of sentences. In order to model the relevance of specific semantics of terms from different aspects, Zhang et al. also introduced the aspect-aware attention mechanism and combined it with the self-attention mechanism. In order to highlight the aspect-related semantic information, Zhang et al. also took the attention score as the initial adjacency matrix of the graph convolutional network, constructed a syntactic mask matrix calculated from the different distances between words with the syntactic information, and used the aspect word mask matrix corresponding to the position of the aspect word to highlight the features after the graph convolutional network. Current research focuses on the simultaneous use of attention mechanisms and GCNs, beginning attempts to combine semantic information obtained using attention mechanisms with grammatical information obtained using GCNs. Therefore, the current research is devoted to finding a more efficient way to combine semantic and grammatical information.

3. The Proposed SSGCN Model

In this section, we propose the sentiment-supported graph convolutional network (SSGCN) model, as shown in Figure 3.

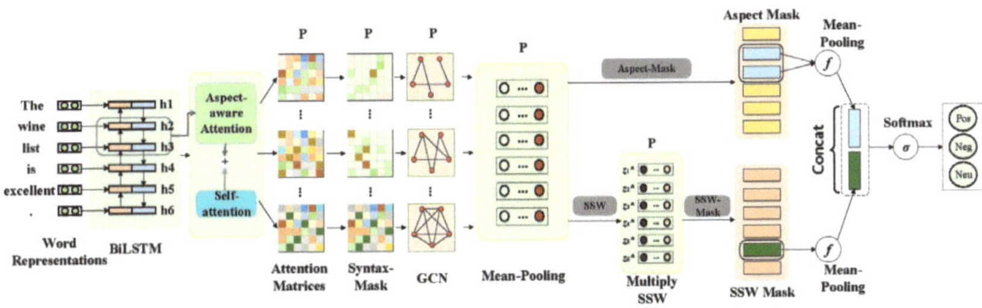

Figure 3. Architecture of the sentiment-supported graph convolutional network (SSGCN) model. Herein, * Represents multiplication.

3.1. Input and Coding Layers

We first used GloVe embedding [30] to obtain the word vector representation of each word, and then we used the Bi-LSTM network to learn the hidden representation H of each word in a sentence. Here, $H = \{h_1, h_2, h_3, \ldots, h_N\}$ contains the hidden representation subsequence $h_a = \{h_{a1}, h_{a2}, h_{a3}, \ldots, h_{am}\}$ corresponding to aspect A.

3.2. Attention Layer

We used an aspect-aware attention mechanism and a self-attention mechanism to form the attention layer. The aspect-aware attention mechanism mainly learns and obtains aspect-related features by calculating the attention of each aspect word. Simultaneously, information regarding the interaction between any two words in a sentence is obtained through the self-attention mechanism. We used the multi-head (P-head in Figure 3) aspect-perceived attention mechanism to obtain the attention score matrix of a sentence, and we also used the P-head self-attention mechanism to obtain a score matrix, finally combining the aspect-perceived attention score with the self-attention score, as shown in Equation (1).

$$A^i = A^i_{asp} + A^i_{self} \tag{1}$$

where A^i_{asp} and A^i_{self} represent the attention scores obtained from the ith aspect-perceived attention head and the ith aspect self-attention head, respectively.

3.3. Syntax Mask Layer

In this study, we used Stanford Parser 3.9.2 (Stanford NLP Group, Stanford, CA, USA) to obtain the syntactic parsing trees. In the syntax mask layer, some nodes in the parsing tree are masked based on different syntactic distances. The model takes the shortest path between nodes V_i and V_j in the parsing tree as the syntactic distance $D(i,j)$ and constructs P syntactic mask matrices based on the syntactic distance. Herein, according to Figure 3, we set $P = 5$. Then, the attention score matrix is combined with the syntactic mask matrix to obtain the syntactic mask matrix A_{mask} using Equation (2).

$$A^i_{mask} = Softmax(A^i + M^i) \tag{2}$$

where A^i_{mask} represents the syntactic mask matrix obtained by combining the ith attention score matrix with the ith syntactic mask matrix, and M^k represents the kth syntactic mask matrix.

3.4. Graph Convolutional Network Layer

Due to there being $P = 5$ different syntactic mask matrices, we performed the graph convolution operation 5 times. For graph convolutional networks, whether the nodes are connected depends upon the syntactic mask matrix A_{mask}. We updated the value of the

ith node in layer l, as shown in Equation (3). The final output of the GCN is shown in Equation (4).

$$h_i^l = \sigma(\sum_{j=1}^{n} A_{ij} W^l h_j^{l-1} + b^l) \quad (3)$$

$$H_l = \left\{ h_1^l, h_2^l, h_3^l, \ldots, h_n^l \right\} \quad (4)$$

where W^l and b^l are denoted as linear transformation weights and bias terms, respectively. σ represents the nonlinear functions. h^{l-1} is denoted as the input of layer l, and h^l represents the output of layer l. H^l indicates the final output of the GCN.

3.5. Semantic and Syntactic Layers of Sentiment Support

In this section, SSWs are used to achieve a preliminary combination of semantics and grammar. We obtained each word's part of speech in a sentence according to the dependency analysis tree, obtaining the semantic information of a sentence from the part of speech perspective. As shown in Figure 4, we conducted the part-of-speech analysis of the aspect words on a widely used dataset, namely, *Laptop*. In its training dataset, we found that among the 2914 pieces of data, 1074 of the aspect words were nouns (NNs), representing 37%. However, the proportion of nouns (NNs) or noun combination phrases (NN-NNs) and plural forms of nouns (NNSs) reached 90%, and the rest were adjectives (JJs) or combinations of adjectives and nouns (JJ-NNs). That is, the aspect terms were mostly noun phrases, nouns, or adjectives plus nouns. Consequently, we proposed three rules for the extraction of emotive support words to determine the emotive support words in specific aspects of sentences from the perspective of grammar, thus acquiring the grammatical information of SSWs in sentences.

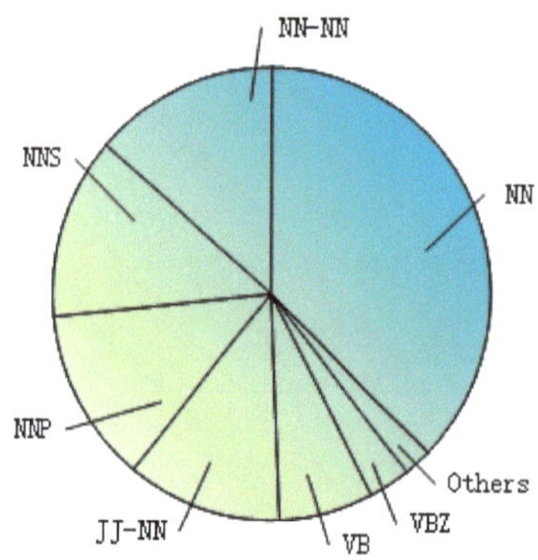

Figure 4. Part of speech statistics for aspect words in the Laptop dataset.

In the syntactic parsing tree, the rules of SSW extraction are as follows:

(1) Noun type: when the part of speech of the aspect word is a noun or another form of a noun (such as a noun singular plural or a proper noun singular plural), we regarded the adjective or other form of the adjective closest to the aspect word in the sentence (such as the adjective comparative or adjective superlative) as the SSW for this aspect.
(2) Combination: when an aspect word is a combination of an adjective or another form of an adjective with a noun or another form of a noun, we considered the adjective or other form of the adjective closest to the aspect word in the sentence as the SSW for that aspect.
(3) Constraint rule: when the relative distance between the adverb and sentiment support word found in the noun and combined types do not exceed two words, we set the adverb as the SSW.

We then used the sentiment support weight z to combine the semantic and grammatical information of the SSW. The weight of the sentiment support is calculated using Equation (5).

$$z = \begin{cases} 2, & w_i \in SSW \\ 1, & \text{otherwise} \end{cases} \quad (5)$$

where z is the sentiment support weight of the word w_i, and SSW is denoted as the sentiment support word. After calculating the sentiment support weight, we used z to weigh the final output representation of the GCN, obtaining the weighted output representation of the GCN H'^{l}.

$$h_i'^{l} = z_i h_i^{l} \quad (6)$$

$$H'^{l} = \left\{ h_1'^{l}, h_2'^{l}, h_3'^{l}, \ldots, h_n'^{l} \right\} \quad (7)$$

3.6. Semantic and Syntactic Information Enhancement Layer

In this section, we further describe the enhancement of the combination of semantic and grammatical information for aspect words with SSWs. Herein, the mask matrix of SSWs is employed to obtain the features of enhanced semantic and grammatical information of SSWs. Similarly, the mask matrix of aspect words is employed to obtain the features from the enhanced semantic and grammatical information of aspect words. For the semantic and grammatical information enhanced for SSWs, we masked the non-sentiment support words in the weighted representation H'^{l} of the GCN output to obtain an SSW representation, retaining important information in the SSW representation through average pooling, as shown in Equation (8).

$$H_{SSW}'^{l} = f(h_{SSW_1}'^{l}, h_{SSW_2}'^{l}, h_{SSW_3}'^{l}, \ldots, h_{SSW_r}'^{l}) \quad (8)$$

For the semantic and grammatical information enhanced for aspect words, we masked the final output of the GCN to represent non-aspect words in H'^{l} to obtain aspect word representations. Similarly, in Equation (9), we retained important information in each aspect word representation through average pooling.

$$H_{asp}'^{l} = f(h_{a_1}'^{l}, h_{a_2}'^{l}, h_{a_3}'^{l}, \ldots, h_{a_r}'^{l}) \quad (9)$$

where f represents the function of average pooling.

We then concatenated the SSW representation with the aspect word representation to obtain a more complete semantic and syntactic final feature representation. Finally, we represented the final feature by obtaining the probability distribution via the *Softmax* function, as shown in Equations (10) and (11).

$$H_{con}'^{l} = Concat(H_{ssw}'^{l}, H_{asp}'^{l}) \quad (10)$$

$$y = Softmax(W_c H_{con}'^{l} + b_c) \quad (11)$$

where *Concat* represents the concatenation operation, and W_c and b_c are linear transform weights and bias terms, respectively.

4. Experimental Analysis

4.1. Datasets

In this study, we used three public datasets to verify our proposed model. The datasets of *Restaurant* and *Laptop* were obtained from SemEval 2014 Task 4 [31], and the third dataset was *Twitter* [32]. Table 1 shows the number of sentences marked with positive, neutral, or negative emotional polarity in the training and test sets of the above datasets.

Table 1. Total number display table.

Dataset	Emotional Polarity	*Restaurant*	*Laptop*	*Twitter*
Training Set	Positive	2164	976	1507
	Neutral	807	851	1528
	Negative	637	455	3016
Test Set	Positive	727	337	172
	Neutral	196	128	169
	Negative	196	167	336

4.2. Parameter Settings

This study employed a 300-dimension pre-trained Glove as a word embedding. The hidden layer dimension of the Bi-LSTM was set to 50. The dropout rate was set to 0.3. The batch size was set to 16 for all models, and the learning rate was set to 0.002. We used Adam as the optimizer.

To ensure the model focused on SSWs, we conducted experiments on the sentiment support weight values of SSWs. We experimented with different weight values in the SSGCN, and the specific experimental results are shown in Table 2. Herein, the accuracy rate and macro F1 were used as evaluation indices.

Table 2. Sentiment support weight value.

z	*Restaurant*	
	Acc	**Macro F1**
1.1	83.65	76.25
1.2	82.57	75.04
1.5	83.02	75.91
1.7	84.18	76.94
2.0	**84.36**	**77.67**
2.2	83.11	75.85
2.5	83.20	74.87

We believe that the weight of the sentiment support should not be too large to highlight the sentiment support words or too small. From Table 2, it can be seen that when the sentiment support weight z was set to 2, using the accuracy and macro F1 as evaluation indicators, the sentiment analysis of the *Restaurant* dataset obtained the best results. Therefore, in this study, we set $z = 2$.

4.3. Baseline Models

In order to fully evaluate the proposed model, we used the following models as baseline methods:

(1) IAN [9]: interactive attention networks are based on LSTM and attention mechanisms.
(2) MGAN [33]: the combined coarse- and fine-grained attention mechanisms are used to create a multi-grained attention network.
(3) MAN [34]: employs intra- and inter-level multi-attention networks.
(4) KumaGCN [35]: a dynamic combination of word dependence graphs and latent infographs learned using self-attention networks.
(5) DualGCN [28]: two GCNs designed to obtain syntactic and semantic information.
(6) CNN-BiLSTM [36]: transfers the data learned from document- to aspect-level sentiment classification.
(7) MIGCN [37]: multi-interaction GCNs that process the semantic and grammatical information between words.
(8) SSEGCN [11]: a syntactic- and semantic-enhanced GCN in which the related feature information of aspect words is enhanced.

4.4. Experimental Results

As shown in Table 3, compared with the baseline models, we found that the accuracy improvement range of the SSGCN on *Restaurant* was 0.54–5.27%, and the improvement range of the macro F1 was 0.75–11.68%. The accuracy improvement range on *Laptop* was from 0.95% to 6.70%. The accuracy improvement range on *Twitter* was 0.10–3.43%, and its improvement range for the macro F1 was 0.32–4.04%. In summary, on all three datasets, the SSGCN showed better accuracy compared with the models based on attention structures or GCNs. In the *Restaurant* and *Twitter* datasets, the evaluated macro F1 results illustrate that the SSGCN was the best model. Our model achieved only a sub-optimal effect in terms of the macro F1 value on the *Laptop* dataset. This was because the syntax structure of the *Laptop* dataset was relatively chaotic, which increased the difficulty of obtaining syntactic information for certain samples. The experimental results demonstrate that our model achieved higher accuracy by incorporating the semantic and grammatical information of aspect words with SSWs.

Table 3. The results of the different comparison models. * indicates the results are directly cited from the original paper (%).

Model	*Restaurant*		*Laptop*		*Twitter*	
	Acc	Macro F1	Acc	Macro F1	Acc	Macro F1
IAN (2017)	79.09	65.99	72.63	67.52	72.50	70.81
MGAN (2018)	81.15	71.64	75.17	70.84	72.54	70.52
MAN (2020)	83.47	70.91	77.21	72.55	75.58	71.15
KumaGCN (2020)	81.43 *	73.64 *	76.12 *	72.42 *	72.45 *	70.77 *
DualGCN (2021)	83.02	76.07	76.58	73.11	75.18	74.06
CNN-BiLSTM (2022)	81.96 *	74.16 *	72.26 *	68.95 *	-	-
MIGCN (2022)	82.32 *	74.31 *	76.59 *	72.44 *	73.31 *	72.12 *
SSEGCN (2022)	83.82	76.92	78.01	**75.35**	73.31	72.12
SSGCN (textual model)	**84.36**	**77.67**	**78.96**	75.17	**75.88**	**74.56**

4.5. Fixed Random Seed

Since parameters in neural networks are randomly initialized by default, setting random seeds can enhance the neural network's fitting ability; however, the results are different each time. To illustrate the effectiveness of our model, we set random number seeds for the CPU, Numpy, and all GPUs to ensure that fixed random numbers were generated each time, ensuring that the experimental results of the same model structure were consistent each time. As shown in Table 4, we set the random number seed to 1000. The results show that our SSGCN was superior to the SSEGCN. Therefore, we argue that the model error in this study was not caused by random disturbances, but rather reflected the stable improvement resulting from model enhancements.

Table 4. Fixed randomized seed experiment results.

Model	*Restaurant*		*Laptop*		*Twitter*	
	Acc	Macro F1	Acc	Macro F1	Acc	Macro F1
SSEGCN	83.47	75.90	77.53	74.25	74.74	72.80
SSGCN	**84.00**	**76.81**	**78.27**	**75.30**	**75.48**	**74.21**

4.6. Comparison of Model Concatenation Methods

We used the concatenation operation to connect the feature representations of aspect words and SSWs, sending the final feature representations to Softmax for emotional classification. To illustrate why Concat was used, we applied the *Restaurant* dataset as an example to compare Concat with direct addition and multiplication operations. The experimental results are shown in Table 5. From this table, it can be seen that using the Concat operation yielded better results than other operations in terms of accuracy and macro F1.

Table 5. Concatenation method results.

Model Concatenation Methods	Acc	Macro F1
Concat	**84.36**	**77.67**
Add	83.47	76.71
Multiply	83.29	75.26

Through the experimental results, it can be seen that the operation of directly adding or multiplying the feature representations of aspect words and sentiment support words resulted in partially lost important information, which was inferior to the concatenation operation with regard to the accuracy and macro F1. Therefore, the concatenation operation was selected for the model in this section.

4.7. Significant Differences between the SSGCN and SSEGCN

Compared with the SSEGCN, the improvements in the proposed model proposed were relatively small. Therefore, in this study, we conducted a *t*-test to determine whether there was a significant difference in the results of our model compared with the SSEGCN, as shown in Table 6.

Table 6. *t*-test results.

Dataset	Acc		Macro F1	
	P	Significance	P	Significance
Restaurant	0.003	Remarkable	0.010	Remarkable
Laptop	0.0001	Remarkable	0.146	Not remarkable
Twitter	0.017	Remarkable	0.037	Remarkable

The null hypothesis was that there was no significant difference between the improved results of the SSGCN and the SSEGCN. The *t*-test showed that when the significance level is 0.05, the null hypothesis can be accepted if p is greater than 0.05; the null hypothesis is rejected if p is less than 0.05. As shown in Table 6, for the *Restaurant* and *Twitter* datasets, there were significant differences between our method and the SSEGCN. There were significant differences in accuracy on the *Laptop* dataset, and the macro F1 value was not significantly improved. The *t*-test generally showed that our model could effectively improve the results.

4.8. Ablation Experiments

In this study, the effectiveness of each module in the SSGCN was checked through ablation experiments. As shown in Table 7, the SSGCN was taken as the benchmark model. Without SSWs denotes that only the weight of the sentiment support was removed, while without SSWMask denotes that only the enhancement of the meaning of sentiment support words and grammatical information was removed. Without AspMask represents the enhancement of only removing aspect word meaning and grammatical information, while without SSWAspMask represents the enhancement of semantic and grammatical information and removing both sentiment support words and aspect words. The following conclusions can be drawn from the results in Table 7:

(1) By comparing without SSWs with the SSGCN, we showed that the sentiment support weight obtained by combining the syntactic and semantic information of SSWs could improve the performance. Specifically, it improved the accuracy by 0.55–1.2% and macro F1 by 0.08–2.23%.
(2) By comparing without SSWMask with the benchmark model, the validity of the enhancement module of the sentiment support word meaning and grammatical information was verified.

(3) By comparing without AspMask with the benchmark model, the validity of the enhancement module of aspect word meaning and syntactic information was verified.
(4) By comparing without SSWMask and without AspMask with the benchmark model, we demonstrated the effectiveness of the feature concatenation module.
(5) According to the comparison between without SSWAspMask and the benchmark model, the accuracy and macro F1 of the model decreased when the semantic and grammatical information of aspect words and SSWs were not enhanced, further indicating the importance and effectiveness of the grammatical and semantic information of aspect words and SSWs.

Table 7. Ablation results.

Model	*Restaurant*		*Laptop*		*Twitter*	
	Acc	Macro F1	Acc	Macro F1	Acc	Macro F1
SSGCN	**84.36**	**77.67**	**78.96**	**75.17**	**75.88**	**74.56**
Without SSWs	83.56	75.46	78.32	74.41	75.33	74.48
Without SSWMask	83.36	76.96	77.06	73.43	75.04	73.43
Without AspMask	83.47	76.31	76.42	72.21	75.78	74.47
Without SSWAspMask	82.84	75.22	78.24	74.45	74.45	73.08

5. Conclusions

In this paper, we present the SSGCN by introducing a graph attention mechanism to obtain semantic and grammatical information about aspect words and SSWs. By constructing the model in this way, the combination of semantic and syntactic information obtained from the SSGCN is more complete. The experimental results illustrate that the accuracy and macro F1 of the proposed model on the open datasets were better than those of the comparison models. More precisely, the improvement in accuracy was 6.33–0.5%. When using macro F1 for evaluation, our method was only 0.18% worse than the SSEGCN, and the improvement range was 11.68–0.5%. We showed that the aspect-level sentiment information was more abundant and the results were more accurate when introducing the semantic and grammatical information of SSWs. However, we searched for SSWs at a relative distance of two words from the aspect word in the syntax analysis tree. This approach is suitable for declarative sentences, but it may not be as effective for other sentence structures. In the future, we will set rules for different sentence structures to improve the accuracy of identifying aspect words.

Author Contributions: Investigation, Y.H.; supervision, Y.H.; writing—original draft, R.G., Z.Z. and Y.L.; writing—review and editing, L.J. All authors read and agreed to the published version of the manuscript.

Funding: This research was funded by the Open Fund Project of National Key Laboratory of Offshore Oil and Gas Development.

Institutional Review Board Statement: Not applicable.

Informed Consent Statement: Not applicable.

Data Availability Statement: The Restaurant and Laptop presented in this study are openly available in SemEval-2014 Task 4 at https://alt.qcri.org/semeval2014/task4/index.php?id=data-and-tools, reference number [28]. The Twitter in this study are openly available in https://goo.gl/5Enpu7, reference number [29].

Conflicts of Interest: The authors declared no potential conflicts of interest with respect to the resaerch.

References

1. Iglesias, C.A.; Moreno, A. Sentiment analysis for social media. *Appl. Sci.* **2019**, *9*, 5037. [CrossRef]
2. Tan, K.L.; Lee, C.P.; Lim, K.M. A survey of sentiment analysis: Approaches, datasets, and future research. *Appl. Sci.* **2023**, *13*, 4550. [CrossRef]

3. Jiang, L.; Li, Y.; Liao, J.; Zou, Z.; Jiang, C. Research on non-dependent aspect-level sentiment analysis. *Knowl.-Based Syst.* **2023**, *266*, 110419. [CrossRef]
4. Moreno, A.; Iglesias, C.A. Understanding Customers' Transport Services with Topic Clustering and Sentiment Analysis. *Appl. Sci.* **2021**, *11*, 10169. [CrossRef]
5. Cioroianu, I.; Corbet, S.; Larkin, C. The differential impact of corporate blockchain-development as conditioned by sentiment and financial desperation. *J. Corp. Financ.* **2021**, *66*, 101814. [CrossRef]
6. Risius, M.; Breidbach, C.F.; Chanson, M.; von Krannichfeldt, R.; Wortmann, F. On the performance of blockchain-based token offerings. *Electron. Mark.* **2023**, *33*, 32. [CrossRef]
7. Wang, Y.; Huang, M.; Zhu, X.; Zhao, L. Attention-based LSTM for aspect-level sentiment classification. In Proceedings of the 2016 Conference on Empirical Methods in Natural Language Processing, Austin, TX, USA, 1–6 November 2016; pp. 606–615.
8. Gers, F.A.; Schraudolph, N.N.; Schmidhuber, J. Learning precise timing with LSTM recurrent networks. *J. Mach. Learn. Res.* **2002**, *3*, 115–143.
9. Ma, D.; Li, S.; Zhang, X.; Wang, H. Interactive Attention Networks for Aspect-Level Sentiment Classification. In Proceedings of the 26th International Joint Conference on Artificial Intelligence, Melbourne, Australia, 19–25 August 2017; pp. 4068–4074.
10. Zhang, C.; Li, Q.; Song, D. Aspect-based Sentiment Classification with Aspect-specific Graph Convolutional Networks. In Proceedings of the 2019 Conference on Empirical Methods in Natural Language Processing and the 9th International Joint Conference on Natural Language Processing, Hong Kong, China, 3–7 November 2019; pp. 4568–4578.
11. Zhang, Z.; Zhou, Z.; Wang, Y. SSEGCN: Syntactic and semantic enhanced graph convolutional network for aspect-based sentiment analysis. In Proceedings of the 2022 Conference of the North American Chapter of the Association for Computational Linguistics: Human Language Technologies, Washington, DC, USA, 10–15 July 2022; pp. 4916–4925.
12. Kim, D.; Kim, Y.J.; Jeong, Y.S. Graph Convolutional Networks with POS Gate for Aspect-Based Sentiment Analysis. *Appl. Sci.* **2022**, *12*, 10134. [CrossRef]
13. Jiang, L.; Zou, Z.; Liao, J.; Li, Y. A Study on the Application of Sentiment-Support Words on Aspect-Based Sentiment Analysis. *Int. J. Pattern Recognit. Artif. Intell.* **2023**, *37*, 2357004. [CrossRef]
14. Sánchez-Rada, J.F.; Iglesias, C.A. Social context in sentiment analysis: Formal definition, overview of current trends and framework for comparison. *Inf. Fusion* **2019**, *52*, 344–356. [CrossRef]
15. Kiritchenko, S.; Zhu, X.; Cherry, C.; Mohammad, S. NRC-Canada-2014: Detecting aspects and sen-timent in customer reviews. In Proceedings of the 8th International Workshop on Semantic Evaluation, Dublin, Ireland, 23–24 August 2014; pp. 437–442.
16. Kim Y. Convolutional Neural Networks for Sentence Classification. In Proceedings of the 2014 Conference on Empirical Methods in Natural Language Processing, Doha, Qatar, 25–29 October 2014; pp. 1746–1751.
17. Wang, J.; Yu, L.C.; Lai, K.R.; Zhang, X. Dimensional sentiment analysis using a regional CNN-LSTM model. In Proceedings of the 54th Annual Meeting of the Association for Computational Linguistics, Berlin, Germany, 7–12 August 2016; pp. 225–230.
18. Liao, S.; Wang, J.; Yu, R.; Sato, K.; Cheng, Z. CNN for situations understanding based on sentiment analysis of twitter data. *Procedia Comput. Sci.* **2017**, *111*, 376–381. [CrossRef]
19. Mutinda, J.; Mwangi, W.; Okeyo, G. Sentiment analysis of text reviews using lexicon-enhanced bert embedding (LeBERT) model with convolutional neural network. *Appl. Sci.* **2023**, *13*, 1445. [CrossRef]
20. Viadinugroho, R.A.A.; Rosadi, D. A Weighted Metric Scalarization Approach for Multiobjective BOHB Hyperparameter Optimization in LSTM Model for Sentiment Analysis. *Inf. Sci.* **2023**, *664*, 119282. [CrossRef]
21. Tang, D.; Qin, B.; Feng, X.; Liu, T. Effective LSTMs for Target-Dependent Sentiment Classification. In Proceedings of the 26th International Conference on Compu-tational Linguistics, Osaka, Japan, 11–16 December 2016; pp. 3298–3307.
22. Gu, T.; Zhao, H.; Li, M. Effective inter-aspect words modeling for aspect-based sentiment analysis. *Appl. Intell.* **2023**, *53*, 4366–4379. [CrossRef]
23. Phan, H.T.; Nguyen, N.T.; Hwang, D. Aspect-level sentiment analysis: A survey of graph convolutional network methods. *Inf. Fusion* **2023**, *91*, 149–172. [CrossRef]
24. Huang, Y.; Liu, Q.; Peng, H.; Wang, J.; Yang, Q. Sentiment classification using bidirectional LSTM-SNP model and attention mechanism. *Expert Syst. Appl.* **2023**, *221*, 119730. [CrossRef]
25. Yadav, R.K.; Jiao, L.; Goodwin, M.; Granmo, O.C. Positionless aspect based sentiment analysis using attention mechanism. *Knowl.-Based Syst.* **2021**, *226*, 107136. [CrossRef]
26. Zhang, S.; Tong, H.; Xu, J.; Maciejewski, R. Graph convolutional networks: A comprehensive review. *Comput. Soc. Netw.* **2019**, *6*, 11. [CrossRef] [PubMed]
27. Yuan, P.; Jiang, L.; Liu, J.; Zhou, D.; Li, P.; Gao, Y. Dual-Level Attention Based on a Heterogeneous Graph Convolution Network for Aspect-Based Sentiment Classification. *Wirel. Commun. Mob. Comput.* **2021**, *2021*, 6625899. [CrossRef]
28. Li, R.; Chen, H.; Feng, F.; Ma, Z.; Wang, X.; Hovy, E. Dual graph con-volutional networks for aspect-based sentiment analysis. In Proceedings of the 59th Annual Meeting of the Association for Computational Linguistics and the 11th International Joint Conference on Natural Language Processing, Online, 11–6 August 2021; pp. 6319–6329.
29. Yang, C.; Song, J.; Yao, S. A Weighted Dependency Tree Convolutional Networks for Aspect-Based Sentiment Analysis. *J. Comput. Eng. Appl.* **2023**, *59*, 129–137.
30. Pennington, J.; Socher, R.; Manning, C.D. Glove: Global vectors for word representation. In Proceedings of the 2014 Conference on Empirical Methods in Natural Language Processing (EMNLP), Doha, Qatar, 25–29 October 2014; pp. 1532–1543.

31. Pontiki, M.; Galanis, D.; Papageorgiou, H.; Androutsopoulos, I.; Manhar, S.; Al-Smadi, M.; Al-Ayyoub, M.; Zhao, Y.; Qin, B.; De Clercq, O.; et al. Semevla-2014 task 4: Aspect based sentiment analysis. In Proceedings of the 8th International Workshop on Semantic Evaluation, Dublin, Ireland, 16–17 July 2014; pp. 27–35.
32. Dong, L.; Wei, F.; Tan, C.; Tang, D.; Zhou, M.; Xu, K. Adaptive recursive neural network for target-dependent twitter sentiment classification. In Proceedings of the 52nd Annual Meeting of the Association for Compu-tational Linguistics, Baltimore, MD, USA, 22–27 June 2014; pp. 49–54.
33. Fan, F.; Feng, Y.; Zhao, D. Multi-grained attention network for aspect-level sentiment classification. In Proceedings of the 2018 Conference on Empirical Methods in Natural Language Processing, Brussels, Belgium, 2–4 November 2018; pp. 3433–3442.
34. Xu, Q.; Zhu, L.; Dai, T.; Yan, C. Aspect-based sentiment classification with multi-attention network. *Neurocomputing* **2020**, *388*, 135–143. [CrossRef]
35. Chen, C.; Teng, Z.; Zhang, Y. Inducing target-specific latent structures for aspect sentiment classification. In Proceedings of the 2020 Conference on Empirical Methods in Natural Language Processing, Online, 8–12 November 2020; pp. 5596–5607.
36. Ayetiran, E.F. Attention-based aspect sentiment classification using enhanced learning through CNN-BiLSTM networks. *Knowl.-Based Syst.* **2022**, *252*, 109409. [CrossRef]
37. Ruyan, W.; Zhongyuan, T.; Rongjian, Z.; Puning, Z.; Zhigang, Y. Multi-interaction Graph Convolutional Networks for Aspect-level Sentiment Analysis. *J. Electron. Inf. Technol.* **2022**, *44*, 1111–1118.

MDPI AG
Grosspeteranlage 5
4052 Basel
Switzerland
Tel.: +41 61 683 77 34

Applied Sciences Editorial Office
E-mail: applsci@mdpi.com
www.mdpi.com/journal/applsci

www.ingramcontent.com/pod-product-compliance
Lightning Source LLC
Chambersburg PA
CBHW041350210926
51850CB00013B/3042
* 9 7 8 3 7 2 5 8 4 5 3 3 0 *